Hanna Seitz

Contributions to the Minimum Linear Arrangement Problem

Hanna Seitz

Contributions to the Minimum Linear Arrangement Problem

On a Binary Distance Model for the Minimum Linear Arrangement Problem

Südwestdeutscher Verlag für Hochschulschriften

Impressum/Imprint (nur für Deutschland/ only for Germany)
Bibliografische Information der Deutschen Nationalbibliothek: Die Deutsche Nationalbibliothek verzeichnet diese Publikation in der Deutschen Nationalbibliografie; detaillierte bibliografische Daten sind im Internet über http://dnb.d-nb.de abrufbar.
Alle in diesem Buch genannten Marken und Produktnamen unterliegen warenzeichen-, marken- oder patentrechtlichem Schutz bzw. sind Warenzeichen oder eingetragene Warenzeichen der jeweiligen Inhaber. Die Wiedergabe von Marken, Produktnamen, Gebrauchsnamen, Handelsnamen, Warenbezeichnungen u.s.w. in diesem Werk berechtigt auch ohne besondere Kennzeichnung nicht zu der Annahme, dass solche Namen im Sinne der Warenzeichen- und Markenschutzgesetzgebung als frei zu betrachten wären und daher von jedermann benutzt werden dürften.

Verlag: Südwestdeutscher Verlag für Hochschulschriften Aktiengesellschaft & Co. KG
Dudweiler Landstr. 99, 66123 Saarbrücken, Deutschland
Telefon +49 681 37 20 271-1, Telefax +49 681 37 20 271-0
Email: info@svh-verlag.de
Zugl.: Heidelberg, Ruprecht-Karls-Universität, Dissertation, 2010

Herstellung in Deutschland:
Schaltungsdienst Lange o.H.G., Berlin
Books on Demand GmbH, Norderstedt
Reha GmbH, Saarbrücken
Amazon Distribution GmbH, Leipzig
ISBN: 978-3-8381-1760-7

Imprint (only for USA, GB)
Bibliographic information published by the Deutsche Nationalbibliothek: The Deutsche Nationalbibliothek lists this publication in the Deutsche Nationalbibliografie; detailed bibliographic data are available in the Internet at http://dnb.d-nb.de.
Any brand names and product names mentioned in this book are subject to trademark, brand or patent protection and are trademarks or registered trademarks of their respective holders. The use of brand names, product names, common names, trade names, product descriptions etc. even without a particular marking in this works is in no way to be construed to mean that such names may be regarded as unrestricted in respect of trademark and brand protection legislation and could thus be used by anyone.

Publisher: Südwestdeutscher Verlag für Hochschulschriften Aktiengesellschaft & Co. KG
Dudweiler Landstr. 99, 66123 Saarbrücken, Germany
Phone +49 681 37 20 271-1, Fax +49 681 37 20 271-0
Email: info@svh-verlag.de

Printed in the U.S.A.
Printed in the U.K. by (see last page)
ISBN: 978-3-8381-1760-7

Copyright © 2010 by the author and Südwestdeutscher Verlag für Hochschulschriften Aktiengesellschaft & Co. KG and licensors
All rights reserved. Saarbrücken 2010

Zusammenfassung

Das *Minimum Linear Arrangement Problem* (MinLA) besteht darin, für einen gewichteten Graphen eine lineare Anordnung der Knoten zu bestimmen, welche die gewichtete Summe der Kantenlängen minimiert. Die vorliegende Arbeit untersucht den Nutzen einer neuen Modelierung im Rahmen eines Branch-and-Cut-and-Price Algorithmus zur optimalen Lösung des MinLA. Den Kern der Modellierung bilden binäre Variablen d_{ijk}, die genau dann den Wert 1 haben, wenn die Knoten i und j in der Permutation die Distanz k haben. Wir präsentieren angepasste Formulierungen für dicht- und dünnbesetzte Graphen und erläutern die Realisierung eines Branch-and-Cut-and-Price Algorithmus'. Desweiteren werden die verschiedenen Varianten des Algorithmus' diskutiert und evaluiert. Zum Studium der theoretischen Aspekte des MinLA leisten wir einen Beitrag mit der Charakterisierung einer Relaxierung des zugehörigen Polyeders.

Abstract

The *Minimum Linear Arrangement problem* (MinLA) consists in finding an ordering of the nodes of a weighted graph, such that the sum of the weighted edge lengths is minimized. We report on the usefulness of a new model within a branch-and-cut-and-price algorithm for solving MinLA problems to optimality. The key idea is to introduce binary variables d_{ijk}, that are equal to 1 if nodes i and j have distance k in the permutation. We present formulations for complete and for sparse graphs and explain the realization of a branch-and-cut-and-price algorithm. Furthermore, its different settings are discussed and evaluated. To the study of the theoretical aspects concerning the MinLA, we contribute a characterization of a relaxation of the corresponding polyeder.

Acknowledgments

Above all I want to thank Prof. G. Reinelt who allowed me to balance research and teaching, a very interesting combination. Furthermore, he was my sounding board when needed and gave me invaluable advice.

Thanks are due also to Prof. G. Bock for writing the report and being very flexible in adapting to the tight schedule.

Marcus Oswald was the best teacher imaginable for on-the-job-training. Furthermore, I am truly thankful for his countless suggestions and unfailing clear explanations.

I am very greatful to have had Thorsten Bonato as my colleague. He was always willing to listen carefully and tireless in helping me to find answers.

Cara Cocking, Markus Speth and Pei Wang provided a very relaxed atmosphere in the office. Thanks for being such nice co-workers.

I am truly thankful for the mentoring of Dirk O. Theis. Working with him was tremendously inspiring and encouraging. I appreciated his competence and feedback as well as his never ending enthusiasm. It made a great difference during all these years.

To Catherine Proux-Wieland a huge "thank you" for the excellent management of all administrative tasks. Her caring delightful nature creates a very pleasant working atmosphere.

Thanks also to Georgios Nikolis and Adrian Dempwolff for keeping the computer system running. Their attention to every detail made working a pleasure.

Without the presence of Karin Tenschert our social life would have been much less exciting. Thanks for being interested, giving emotional support as well as sharing personal perspectives.

On numerous occasions I experienced great cooperation between various groups and members of the institute. In particular I want to thank Prof. K. Ambos-Spies, Prof. T. Ludwig and Prof. B. Paech for their personal engagement.

Even though and much to my regret, we did not spend much time together, I want to thank Prof. E. Fernandez, Barcelona, for interesting scientific discussions.

I thank Prof. A. Letchford, Lancaster (UK), for our inspiring discussions and exceptionally pleasant teamwork. It has been a great honor to do research with him.

For the proofreading parts of my thesis I thank Thorsten Bonato, Elsbeth Duke, Sarah Maidstone, Marcus Oswald, Robert Schwarz, Christoph Schwitzer, Markus Speth, Dirk O. Theis, and Stefan Wiesberg. Nevertheless, I take full responsibility for any errors that may remain.

Over all these years my family and a lot of friends influenced me enormously by their encouragement and support. I am truly indebted to all minor and major contributions.

Last but not least my heartfelt thanks go to my parents Heike and Friedhelm Peters to whom I am endebted the most.

Contents

Introduction — 1

0 Preliminaries — 3
- 0.1 Graphs — 3
- 0.2 Complexity — 3
- 0.3 Linear Algebra — 4
  - 0.3.1 Matrices — 5
  - 0.3.2 Permutations — 5
  - 0.3.3 Metrics — 5
- 0.4 Polyhedra — 6
- 0.5 Linear Programming — 7
- 0.6 Solution Methods for Linear Programs — 8
  - 0.6.1 Simplex Algorithm — 9
  - 0.6.2 Pricing — 9
  - 0.6.3 Pricing Versus Column Generation — 10
- 0.7 Integer Programming — 11
- 0.8 Solution Methods for Integer Programs — 11
  - 0.8.1 Branch-and-Bound — 11
  - 0.8.2 Cutting Planes — 12
  - 0.8.3 Branch-and-Cut — 13
  - 0.8.4 Branch-and-Cut-and-Price — 13

1 A Brief Survey — 19
- 1.1 Definition of MinLA and Basic Properties — 19
- 1.2 Applications — 20
- 1.3 Complexity — 20
  - 1.3.1 $\mathcal{NP}$-hard Cases — 20
  - 1.3.2 Polynomially Solvable Cases — 20
- 1.4 Computation of Upper Bounds — 22
  - 1.4.1 Heuristics — 22
  - 1.4.2 Approximations — 22
- 1.5 Computation of Lower Bounds — 24
  - 1.5.1 Combinatorial Bounds — 24
  - 1.5.2 Linear Programming Bounds — 25
  - 1.5.3 Other Bounds — 29

2 Binary Distance Model — 31
2.1 Binary Distance Model — 31
2.1.1 Definition and Basic Properties — 31
2.1.2 Integer Programming Formulation — 32
2.1.3 Further Inequalities — 34
2.1.4 Rank Constraints on y-Variables — 37
2.2 Sparse Problem Formulation — 40
2.2.1 Basic Properties and Modified Constraint System — 40
2.2.2 Additional Variables — 40
2.2.3 Shortest Path Strengthening — 41
2.2.4 Extended Star Constraint — 41
2.3 Assignment Variables Formulation Revisited — 42
2.3.1 Improvement of the Formulation — 42

3 Polyhedral Theory — 43
3.1 Preparative Definitions — 43
3.1.1 Pairwise absolute value mapping M — 46
3.2 Integral Polyhedron P_n — 48
3.2.1 Literature Review — 48
3.2.2 Feasibility Test for $P_n(G)$ — 49
3.3 The Convex Set Q_n — 53
3.3.1 Definition of Q_n and Basic Properties — 53
3.3.2 Unbounded Edges of Q_n — 55
3.3.3 The Minkowski Sum $P_n + \text{CUT}_n$ equals $\overline{Q_n}$ — 56
3.3.4 Inequalities Defining Facets of $\overline{Q_n}$ — 66
3.4 0/1 Polytope D_n — 68
3.4.1 Definition of D_n and Basic Properties — 68
3.4.2 Conjectured Minimal Equation System — 68
3.4.3 Small Polytopes — 70
3.5 Integral Distance and Assignment Polyhedron P_n^A — 71
3.5.1 Definition of P_n^A and Basic Properties — 71
3.5.2 Bounding Inequalities — 71
3.5.3 Strengthening the Inequalities from Amaral and Letchford — 73

4 Branch-and-Cut-and-Price Algorithm — 75
4.1 Choice of Start Variables — 75
4.2 Start Heuristics — 75
4.2.1 Simulated Annealing — 75
4.2.2 Multi-Start Local Search Routine — 75
4.3 Separation — 77
4.3.1 Exact Separation Algorithms — 77
4.3.2 Heuristic Separation Algorithms — 78
4.3.3 Separation Speed Up — 79
4.3.4 Cut Selection Strategies — 80
4.4 Improvement Heuristics — 81
4.4.1 Distance to the Border Heuristic — 81
4.4.2 Longest Distances Heuristic — 82
4.4.3 Edge Lengths Heuristic — 82

4.5 Branching . . . 83
4.5.1 Branch on Variables . . . 83
4.5.2 Branch on Constraints . . . 83
4.5.3 Branch on Variables and Constraints . . . 85
4.5.4 Choice of Branch Variable/Constraint . . . 86

5 Computational Results 89
5.1 Test Problem Instances . . . 89
5.2 Computational Setup and Details of the Implementation . . . 90
5.2.1 Memory Management of Variables due to Branch-and-Price . . . 90
5.2.2 Central Saved Adjacency List . . . 90
5.2.3 `SetByLogImp()` . . . 91
5.3 Justification of the 0/1 Model . . . 91
5.4 Complete 0/1 Model . . . 93
5.4.1 Explicit Use of y-Variables . . . 93
5.4.2 LP Solver Settings . . . 93
5.5 Pricing for the 0/1 Model . . . 95
5.5.1 Lower Bound Decrease Due to Pricing . . . 95
5.5.2 Start Sets of Variables . . . 95
5.5.3 Size of the Variable's Start Set . . . 97
5.5.4 Maximal Number of Priced-in Variables per Call . . . 97
5.5.5 Random versus Smart Pricing . . . 97
5.5.6 Additional Pricing Steps . . . 98
5.6 Constraints . . . 99
5.6.1 Strength of Constraint Types . . . 99
5.6.2 Exact versus Heuristic Separation of the Triangle Inequalities . . . 108
5.6.3 Different Subgraph Sizes . . . 108
5.6.4 Combination of Best Constraints . . . 109
5.7 Separation . . . 109
5.7.1 Random versus Smart Separation . . . 109
5.7.2 Different Maximal Numbers of Separated Constraints . . . 109
5.7.3 Cut Selection Strategies . . . 110
5.7.4 Deep Separation Variants . . . 112
5.8 Improvement Heuristics . . . 113
5.9 Branching . . . 114
5.9.1 Small Test Instances . . . 114
5.9.2 Branch on Deg-Big Constraints . . . 114
5.10 Sparse 0/1 Model . . . 115
5.10.1 Complete versus Sparse Problem Formulation . . . 117
5.11 Final Comparisons . . . 119

A Unbounded Edges of Q_n 123

List of Algorithms 125

List of Figures 127

List of Tables 129

References 131

Nomenclature 137

INTRODUCTION

The Minimum Linear Arrangement problem (MinLA) consists in finding an ordering of the nodes of a weighted graph on n nodes, such that the sum of the weighted edge lengths is minimized. This thesis makes a contribution to the computation of lower bounds for the MinLA problem. To do so, we introduce a new modeling based on binary variables and investigate its usefulness within a branch-and-cut-and-price algorithm.

As there are several applications of the MinLA, a lot of work on heuristic solutions of the MinLA can be found [1–4]. In contrast to these approaches, we are interested in finding a provably optimal solution of the MinLA. Compared to many other $\mathcal{NP}$-hard optimization problems, the MinLA turns out to be more difficult and is extremely hard to solve in practice.

To our knowledge the first known suitable integer programming model for the MinLA was formulated by Even et al. [5]. Although no practical results were computed, this sparse formulation was important, as its non-trivial linear programming relaxation can be solved in polynomial time. Another integer programming model for the MinLA was presented by Liu & Vannelli [6]. It is based on rank constraints for suitable subgraphs that can be computed in polynomial time. In contrast to the formulation in [5] mentioned above, this problem formulation is dense, i.e., it is based on *all* distance variables y_{ij} for $i,j \in V$. Caprara et al. [7] based their recent work on both modeling approaches. The advantages of the sparse spreading metric formulation and the dense rank constraint approach were combined by Caprara et al. [7] in a successful way. Their computational results showed that for most benchmark instances, the best-known solutions were not far from the optimum. A different approach is outlined by Caprara et al. [8]. Their key idea was to use betweenness variables for a "refinement" of the integral distance variables y. Although this approach was based on a large number of (n^3) variables, the computational results were comparable to those of [7].

In this study we address the problem by offering a new solution strategy for the MinLA, combining the established and well-working approach of a branch-and-cut algorithm with pricing. We use the distance variables y that were part of the models developed so far, in the following way: We "refine" the variables in such a way that the resulting model is significantly enriched. We accomplish this by introducing our binary distance variables d_{ijk}, where d_{ijk} equals 1 if nodes i and j have distance k.

In the first chapter of this thesis, we start with a literature review of the work done on the MinLA. An overview of some basic properties is given, and several applications and approximation algorithms are presented. Furthermore, all modeling approaches and other lower bounds of the MinLA known to the author are explained.

In Chapter 2 we introduce the binary distance model in its complete and sparse version. We start with the linear inequalities needed for the integer programming formulation and characterize the new model with respect to the similarity and in contrast to the integral y-variables formulation. Further inequalities are presented, and their strength is compared to corresponding y-constraints. The model is investigated in a second step, where we consider only those variables for which an edge in the graph exists. We show how the system of constraints must be modified. As we want to obtain a similar strength of the formulation as in the complete case, we present two different approaches which help to improve the quality of the formulation. We then consider an improved mixed linear programming formulation of the MinLA that uses the integral distance variables y in combination with n^2 binary assignment variables.

The theoretical aspects of the MinLA are presented in Chapter 3. First, we give a short overview of the work on

$P_n(G)$, which is the polyhedron corresponding to the y-variables formulation, and on DOM_n, its Minkowski sum with the non-negative orthant of $\mathbb{R}^{|E|}$. We present a feasibility test for the integral distance formulation and specify a graph property for which the presented algorithm has polynomial running time. We then turn our attention to the convex set Q_n, which is an alternative relaxation of P_n. We are particularly interested in Q_n's relationship to P_n and to the cut cone CUT_n. Furthermore, we characterize its unbounded edges along with the unbounded edges of its closure $\overline{Q_n}$. The polyhedron D_n, corresponding to the complete problem formulation with the d-variables, is investigated in Section 3.4. This is followed by a study of the polyhedron P_n^A of the revisited assignment variables formulation.

Finally, in Chapter 5, computational results are presented and discussed. Before we start with a detailed test of different settings of the algorithm, we present a comparison between the distance and the binary distance modeling approach. We then test the explicit versus the implicit use of y-variables in our branch-and-cut-and-price algorithm and compare the results for different linear programming solver settings to each other. This is followed by a discussion of various start heuristics, which have a great effect due to the generation of the start variables of our algorithm being dependent on the start solution. We continue with a report on the effects of different sizes of the set of start variables. Our investigation of the influence of pricing mechanisms on the performance of the algorithm also includes the use of different pricing strategies and additional pricing steps. Provided with the best choice of all these pricing possibilities, we then test and evaluate the strength of all constraint types. Furthermore, we compare various general separation strategies, such as different modifications of the current LP solution. We continue with a presentation of cut selection strategies such as rankings and variable disjoint cut selection. The usefulness of different improvement heuristics is considered and several branching criteria are tested. Furthermore, computational results for the sparse problem formulation with all additional features are presented. Ultimately we compare our best results with those obtained from using other models of the MinLA.

Chapter 0

Preliminaries

In this chapter, a brief summary of the basic definitions and notations used within this thesis is given. We assume familiarity of the reader with the underlying concepts of mathematics and computer science. For further reading we refer to suitable books in each section.

0.1 Graphs

An undirected (unweighted) **graph** $G = (V,E)$ is a pair of finite sets V and E. The set V contains all **vertices** of G, and $E \subseteq V \times V$ is the set of all **edges** $e = ij := \{i,j\}$ of G. The graph G is **directed** if all pairs of nodes are ordered, i.e., $e = ij \neq ji = e'$. We will denote the number of nodes by $n = |V|$; $m = |E|$ will be the number of edges in G. A node i and an edge e are **incident** if $e = ij$ or $e = ji$. Two edges are **adjacent** if they share a node. Similarly, two nodes are adjacent if there exists an edge which is incident to both.

For each edge e a real number $c_e \in \mathbb{R}_+ \cup \{0\}$ can be assigned which is the **weight** of the edge. If all edges have weights, G is called a **weighted graph**. The weight of a subset $F \subseteq E$ is defined by $c(F) := \sum_{e \in F} c_e$.

Given a subset $S \subseteq V$ of nodes, the **cut** $\delta(S)$ of S is defined by

$$\delta(S) := \{ij \in E \mid i \in S, j \in V \setminus S\}.$$

The **cut matrix** D_S corresponding to S is an $n \times n$ matrix and defined as $(D_S)_{kl} := 1$ if $kl \in \delta(S)$ and otherwise $(D_S)_{kl} = 0$.

The **degree** $\deg(i)$ of a node i is the number of edges incident to i. For undirected graphs we have $\deg(i) = |\delta(\{i\})|$. For a directed graph we denote by $\deg_{\text{in}}(i)$ the edges towards i and with $\deg_{\text{out}}(i)$ the edges from i.

For a subset $F \subseteq E$ of edges the **incidence** or **characteristic vector** $\chi^F \in \{0,1\}^{|E|}$ is defined as

$$\chi^F_e = \begin{cases} 1, & e \in F, \\ 0, & \text{otherwise.} \end{cases}$$

Analogously, the characteristic vector of a node subset $S \subseteq V$ is defined. The superscript "c" denotes the complement of the set.

For a more detailed presentation of these definitions and graph theory we refer the reader to [9, 10] and [11].

0.2 Complexity

This section is a summary of the presentation in [12], the literal excerpt is characterized in quotations.

"A polynomial-time algorithm is an algorithm that terminates after a number of steps bounded by a polynomial in the input size. Here a step consists of performing one instruction. (...) We say a problem is **polynomial-time** solvable, or is solvable in polynomial time, if it can be solved by a polynomial-time algorithm. (...) The collection of all polynomial-time solvable problems (...) is denoted by $\mathcal{P}$." The class $\mathcal{NP}$ (**nondeterministic polynomial-time**) is the collection of decision problems that can be reduced in polynomial time to the satisfiability problem. "Roughly spoken, $\mathcal{NP}$ is defined as the collection of all decision problems for which each input with positive answer has a polynomial-time checkable (...) correctness of the answer. (...) The class $\mathcal{NP}$ is apparently much larger than the class $\mathcal{P}$, and there might be not much reason to believe that the two classes are the same. But, as yet, nobody has been able to prove that they really are different. This is an intriguing mathematical question, but besides, answering the question might also have practical significance. If $\mathcal{P} = \mathcal{NP}$ can be shown, the proof might contain a revolutionary new algorithm, or alternatively it might imply that the concept of 'polynomial-time' is completely meaningless. If $\mathcal{P} \neq \mathcal{NP}$ can be shown, the proof might give us more insight in the reasons why certain problems are more difficult than others and might guide us to detect and attack the kernel of the difficulties."

The hardest problems in $\mathcal{NP}$ are called $\mathcal{NP}$-**complete**: every problem in $\mathcal{NP}$ can polynomially be reduced to them. An optimization problem is called $\mathcal{NP}$-**hard** if and only if a $\mathcal{NP}$-complete problem can be reduced to it in polynomial time.

"Several prominent combinatorial optimization problems, like the Traveling Salesman problem, (...) and the Maximum Cut problem, are $\mathcal{NP}$-hard. $\mathcal{P}$ and $\mathcal{NP}$ are collections of **decision problems**: problems that can be answered by 'yes' or 'no' (...). An optimization problem is no decision problem, but often can be reduced to it in a certain sense. (...) Consider a minimization problem: minimize $f(x)$ over $x \in X$, where X is a collection of elements derived from the input of the problem, and where $f(x)$ is a rational-valued function on X. (...) This can be transformed to the following decision problem: 'given a number r, is there an $x \in X$ with $f(x) \leq r$?' (...) About all combinatorial optimization problems, when framed as a decision problem (...), belong to $\mathcal{NP}$, since a positive answer to the question can often be certified by just specifying an $x \in X$ satisfying $f(x) \leq r$."

0.3 Linear Algebra

For a real vector space L^n of dimension n, we denote by $(L^n)^*$ its dual space. Note that L^n is not the n-fold Cartesian product $L \times L \times \ldots \times L$. We consider the elements of $\mathbb{R}^n$ as column vectors, and for $x \in \mathbb{R}^n$, the transpose x^T is in $(\mathbb{R}^n)^*$. We denote by **1** the appropriately sized vector consisting of ones. Analogously **0** is the vector whose entries are all zero. If appropriate, we will use a subscript $\mathbf{1}_k$, $\mathbf{0}_k$ to identify the lengths of the vectors. The kth unit vector is denoted by e_k. We will abbreviate the set $\{1, \ldots, n\}$ by $[n]$.

The vectors $x_1, \ldots, x_k \in L^n$ are **affinely independent** if the unique solution of the linear system $\sum_{i=1}^{k} \lambda_i x_i = 0$ and $\sum_{i=1}^{k} \lambda_i = 0$, is $\lambda_i = 0$ for all $i = 1, \ldots, k$. Linear independence implies affine independence; the converse, however, is not true. Note: the maximum number of affinely independent points in L^n is $n + 1$. In particular, any n linearly independent points in L^n and the **0** vector are affinely independent.

Given vectors $x_1, \ldots, x_k \in L^n$ we call $\sum_{i=1}^{k} \lambda_i x_i$ a **conic combination** if $\lambda_i \geq 0$ for all $i = 1, \ldots, k$. If instead $\sum_{i=1}^{k} \lambda_i = 1$, we call it an **affine combination**. If a combination is conic and affine, we call it **convex combination**.

The **affine hull** of a set $X \subset L^n$, $X = \{x_1, \ldots, x_k\}$, is the smallest set containing all affine combinations of $x_1, \ldots, x_k$ and denoted by aff$(\{x_1, \ldots, x_k\})$. The **convex hull** conv$(\{x_1, \ldots, x_k\})$ is defined as the smallest convex set containing $x_1, \ldots, x_k$.

A subset $C \subseteq L^n$ is a **cone** if it is closed under conic combination, i.e., if $x, y \in C$ then $\lambda x + \mu y \in C$ for all $\lambda, \mu \in \mathbb{R}_+$.

0.3. Linear Algebra

A cone is **polyhedral** if there exists a finite set of vectors $x_1, \ldots, x_k$ such that
$$C := \left\{ y \in L^n \;\middle|\; y = \sum_{i=1}^{k} \lambda_i x_i, \; \lambda_i \geq 0 \text{ for all } i = 1, \ldots, k \right\}.$$
Within this thesis all cones will be polyhedral, therefore we will omit the explicit specification from now on. Note: any cone must contain the origin and can be represented as $C := \{ x \in L^n \mid Ax \leq 0 \}$, for a suitable matrix A.

0.3.1 Matrices

The symbol $\mathbb{0}$ denotes an all-zeros matrix which is not necessarily square. We also use it to say "this part of the matrix consists of zeros only". By $\mathbb{1}_n$ we denote the square matrix of order n whose (k,l)-entry is 1 if $k \neq l$ and 0 otherwise. We will omit the index n when appropriate. Given two matrices $A, B \in \mathbb{R}^{n \times m}$ we define
$$A \bullet B := \operatorname{tr}(A^T B) = \sum_{k=1}^{n} \sum_{l=1}^{m} A_{kl} B_{kl}.$$
The set of all $n \times n$ matrices with entries in $\mathbb{R}$ is $\mathbb{M}(n, \mathbb{R})$. Moreover the set of all symmetric $n \times n$ matrices with zero in the main diagonal will be $S\!M(n)$. Because of the 0-entries and the symmetry we have $S\!M(n) \cong \mathbb{R}^{\binom{n}{2}}$.

0.3.2 Permutations

The set of all permutations π of n is defined by $S(n)$. We occasionally view $S(n)$ as a subset of $\mathbb{R}^n$ by identifying the permutation π with the point $(\pi(1), \ldots, \pi(n))^T$. With ι_n we denote the identical permutation on $[n]$. As above we omit the index n when no confusion can arise. The so-called **antipodal permutation** π^- is defined by
$$\pi^- := (n+1)\mathbf{1} - \pi,$$
e.g. if $\pi = (2\,3\,1\,4)$ then $\pi^- = (3\,2\,4\,1)$. The **permutation matrix** will be denoted by $E_\pi := (e_{\pi(1)}, \ldots, e_{\pi(n)})^T$. Note that $e_k = E_\pi e_{\pi(k)}$ and $E_\pi^T = E_\pi^{-1} = E_{\pi^{-1}}$.

0.3.3 Metrics

A **semi-metric** on $[n]$ is a mapping $d : [n] \times [n] \to \mathbb{R}_+$ which

- satisfies the triangle inequality,
- is symmetric, i.e., $d(i,j) = d(j,i)$ for all $i, j \in [n]$ and
- satisfies $d(i,i) = 0$ for all $i \in [n]$.

If in addition $d(i,j) = 0$ only if $i = j$ for all $i, j \in [n]$, d is called a **metric**. We understand a metric both as a function and a matrix, and we will switch between the two concepts without further mentioning.

We say a metric d is **embeddable in the real line**, for short d is ℓ_1-embeddable in $\mathbb{R}$, if there exist $x_1, \ldots, x_n \in \mathbb{R}$ with $d_{kl} = |x_k - x_l|$ for all $k, l \in [n]$. It is known that the set of ℓ_1-embeddable semi-metrics on $[n]$ is a polyhedral cone in $\mathbb{R}^{\binom{n}{2}}$ [13]. In fact, it is nothing but the well-known **cut cone**, denoted by CUT_n.

A **cut metric** is defined as follows: For a set $U \subset [n]$, we let d_U be the metric which assigns two points on different sides of the bipartition U, U^c of $[n]$ a value of 1 and two points on the same side a value of 0. We will say that the set U **induces** the associated cut metric. With this notation, CUT_n is the convex cone with apex 0 in $S\!M n$ generated by the points d_U, i.e.,
$$\mathrm{CUT}_n := \operatorname{cone}\left\{ d_U \;\middle|\; d_U \text{ is the cut metric for } U \subset [n] \right\}.$$
It is known that each cut metric defines an extreme ray of CUT_n [14].

We study the metrics d on $[n]$ that arise when n points are embedded in the real line, in such a way that the distance between each pair of points is at least 1. We call these metrics $\mathbb{R}$-**embeddable 1-separated metrics**. We remark that one could easily replace the value 1 with some arbitrary constant $\epsilon > 0$; the results would remain essentially unchanged.

For a deeper insight in linear algebra we refer the reader to [15] and [16].

0.4 Polyhedra

Let L^n be a linear vector space over $\mathbb{R}$. A **polyhedron** $P \subseteq L^n$ is the intersection of finitely many closed half spaces, or equivalently, the solution set of a finite system of linear inequalities, i. e., $P := \{x \in L^n \mid Ax \leq b\}$, where $A \in \mathbb{R}^{m \times n}$ and $b \in \mathbb{R}^m$. This representation of P via the system of inequalities (A,b) is called the **H-representation** or **outer description** of P. For another characterization of polyhedra we need the **Minkowski sum** of two polyhedra P_1 and P_2, which is the set
$$P_1 + P_2 := \{x + y \mid x \in P_1, y \in P_2\}.$$
The so-called **V-representation** or **inner description** of a polyhedron P is $P := \operatorname{conv}(X) + \operatorname{cone}(Y)$, where $X \in \mathbb{R}^{n \times m}$ and $Y \in \mathbb{R}^{n \times m'}$. These two descriptions are equivalent due to

Theorem 0.4.1 ([17]). *A subset $P \subseteq L^n$ is a Minkowski sum of a convex hull of a finite set of points plus a conic hull of a finite set of vectors*
$$P := \operatorname{conv}(X) + \operatorname{cone}(Y), \text{ where } X \in \mathbb{R}^{n \times m}, Y \in \mathbb{R}^{n \times m'}$$
if and only if it is an intersection of closed half spaces
$$P := \{x \in L^n \mid Ax \leq b\}, \text{ for some } A \in \mathbb{R}^{r \times n}, b \in \mathbb{R}^r.$$

If P is bounded we call it a **polytope**. The notion of affine independence is useful in defining the dimension of a polyhedron. The **dimension of a polyhedron** $P = \{x \in L^n \mid Ax \leq b\}$ is k, denoted by $\dim(P) = k$, if the maximum number of affinely independent points in P is $k + 1$. We set $\dim(\varnothing) = -1$.

Proposition 0.4.2 ([18]). *Let $P = \{x \in L^n \mid Ax \leq b\}$ and let A', b' be the subsystem of A, b such that $A'x = b'$ for all solutions of $Ax \leq b$. Then $\dim(P) = n - r$, where r is the rank of A'.*

Let $P \subseteq L^n$ be a polyhedron. An inequality $ax \leq b$, is a **valid inequality** for P if it is satisfied by all elements of P. A **face** of P is the set $F = P \cap \{x \in L^n \mid ax = b\}$ with $ax \leq b$ being a valid inequality of P. As $\mathbf{0}^T x \leq 0$, we get that P itself is a face of P. For the inequality $\mathbf{0}^T x \leq 1$, we see that $\varnothing$ is always a face of P. The **dimension of a face** is the dimension of its affine hull, i. e., $\dim(F) := \dim(\operatorname{aff}(F))$. A face of dimension 0 is called a **vertex** of P, a face of dimension 1 is an **edge**. A **facet** F is a face of P of dimension $\dim(P) - 1$. If $F = P \cap \{x \in L^n \mid ax = b\}$ is a facet, the inequality $ax \leq b$ is called **facet-defining**.

Besides the above descriptions of a polyhedron P, it can also be described in terms of points and rays. To do so we start with some more definitions. A subset X of a convex set C is called **exposed** if there exists a half space H containing C, such that the intersection of the bounding hyperplane of H with C is equal to X. In other words, X is exposed if there exists a valid inequality for C such that X is the set of all points in C satisfying the inequality with equation. A subset X of a convex set C is called **extreme** if $tc + (1 - t)c' \in X$ for $c, c' \in C$ and $0 < t < 1$ implies $c, c' \in X$. Clearly, if X is exposed it is also extreme, but the converse is not necessarily true. It is true if C is closed, which is the case if C is a polyhedron. In this case every exposed set is a face of the polyhedron. Given $x \in P = \{x \in \mathbb{R}^n \mid Ax \leq b\}$ we say x is an **extreme point** of P if there do not exist $x_1, x_2 \in P$, $x_1 \neq x_2$, such that $x = \frac{1}{2}x_1 + \frac{1}{2}x_2$. It is true that x is an extreme point of P if and only if x is a zero dimensional face of P. As the zero dimensional faces are called vertices, we denote the **set of all extreme points** of P by $\operatorname{vert}(P)$. A vector $r \in \mathbb{R}^n$, $r \neq \mathbf{0}$ induces a **ray** $\mathbb{R}_+ r$ of P if $x \in P$ implies $x + \lambda r \in P$ for all $\lambda \in \mathbb{R}_+$. Often we omit to specify that the vector r induces a ray, but call r a ray itself. A ray $\mathbb{R}_+ r$ is an **extreme ray** of P if $\{\lambda r \mid \lambda \in \mathbb{R}_+\}$ is a one dimensional face of $\{r \in \mathbb{R}^n \mid Ar \leq 0\}$. The **set of all extreme rays** of P will be denoted by $\operatorname{exray}(P)$.

Theorem 0.4.3 ([19]). *Let $x_1, \ldots, x_k$ be the extreme points of the nonempty polyhedron $P = \{x \in \mathbb{R}^n \mid Ax \leq b\}$, where $\operatorname{rank}(A) = n$. Further let $r_1, \ldots, r_l$ be its extreme rays. Then*
$$P = \left\{ x \mid x = \sum_{i=1}^k \lambda_i x_i + \sum_{j=1}^l \mu_j r_j, \text{ where } \mathbf{1}^T \lambda = 1, \lambda \in (\mathbb{R}_+^n)^k, \mu \in (\mathbb{R}_+^n)^l \right\}.$$

The **recession cone** $\text{rec}(P)$ of a polyhedron P is the set of all infinite directions in P, i.e.,

$$\text{rec}(P) := \{y \in L^n | x + ty \in P \text{ for all } x \in P, t \geq 0\}.$$

We set $\text{rec}(\varnothing) := \{\mathbf{0}\}$.

Theorem 0.4.4 ([17]). *A polyhedron P can be described with the recession cone via the following equation*

$$P = \text{rec}(P) + \text{conv}\{x | x \in \text{vert}(P)\}.$$

Let P be a polytope in L^n and $y \in P$. If y is not contained in a face of P of dimension smaller than n, it is an **interior point** of the polytope. Moreover, the **set of all interior points** of P is denoted by $\text{int}(P)$. It can easily be shown that this definition coincides with the usual topological definition of the interior of the set $P \subseteq L^n$. Note: the interior of a polytope is not invariant under affine equivalence of polytopes and that, in fact, $\text{int}(P) = \varnothing$ if P is not full-dimensional in L^n.

A polyhedron P is called **simple**, if every vertex of P is contained in exactly $\dim(P)$ facets. It is **simplicial** if and only if all facets contain only $\dim(P)$ vertices.

The **polar** $P^\triangle$ of a polyhedron $P \subseteq L^n$ is the set of all left hand sides of normalized valid inequalities:

$$P^\triangle := \left\{ c \in (L^n)^* \mid c^T x \leq 1 \text{ for all } x \in P \right\}.$$

It can only be defined correctly if $\mathbf{0} \in \text{int}(P)$. The vertices in P correspond exactly to the facets in $P^\triangle$ and vice versa. Given two polyhedra P_1 and P_2 where $P_1 \subseteq P_2$, we have $P_1^\triangle \supseteq P_2^\triangle$. If a polyhedron is simple, its polar is simplicial (and vice versa).

Within this thesis we will not only work with polyhedra in the real vector space $\mathbb{R}^n$, but with polyhedra in different linear vector spaces, e.g. $L^n = \mathbb{SM}(n)$. For more information about these definitions we refer to [17–19] and [12].

0.5 Linear Programming

The following definitions are based on the presentation in [20]. The **linear programming problem** consists of finding a vector $x \in \mathbb{R}^n$ that fulfills all given **constraints** $Ax \leq b$ and maximizes a certain **objective function** $c^T x$, where A is an $m \times n$ matrix and $b \in \mathbb{R}^m$ and $c \in \mathbb{R}^n$ are vectors. We denote this problem as **linear program** (LP) or **primal program**, $\mathcal{PP}$ for short. Its **standard form** is the following:

$$(\mathcal{PP}) \quad \begin{aligned} \min \quad & c^T x \\ \text{s.t.} \quad & Ax \leq b \\ & x \in \mathbb{R}^n. \end{aligned}$$

A vector $x \in \mathbb{R}^n$ that satisfies $Ax \leq b$ is called a **feasible solution**. A feasible solution that is maximal is called an **optimal solution**. Each linear program can be associated with a so-called **dual program**

$$(\mathcal{DP}) \quad \begin{aligned} \max \quad & y^T b \\ \text{s.t.} \quad & y^T A = c^T \\ & y \leq \mathbf{0}, \end{aligned}$$

where y is the variable vector. The linear program is often called the **primal program**. Table 0.1 shows how the two programs $\mathcal{PP}$ and $\mathcal{DP}$ can be transformed into one another. Here $A_{\cdot j}$ is the jth column and $A_{j \cdot}$ the jth row of matrix A.

The following important **duality theorem** establishes an important connection between $\mathcal{PP}$ and $\mathcal{DP}$.

Theorem 0.5.1 ([20]). *(a) If $\mathcal{PP}$ and $\mathcal{DP}$ both have feasible solutions, then both problems have optimal solutions and the optimum values of the objective functions are equal.*

Table 0.1: Correspondence of the primal and dual linear program.

In the dual ($\mathcal{DP}$)	In the primal ($\mathcal{PP}$)
max	min
variables x_j, $j = 1,\ldots,n$	functions $A_{j.}^T u$, $j = 1,\ldots,n$
functions $A_{.j}^T x$, $j = 1,\ldots,n$	variables x_j, $j = 1,\ldots,n$
objective function $c^T x$	right hand side c
right hand side b	objective function $b^T u$
constraints $A_{l.} x \leq b_l$ for all $l \in L$	variables $u_l \geq 0$ for all $l \in L$
$A_{e.} x = b_e$ for all $e \in E$	u_e free for all $e \in E$
$A_{g.}^T x \geq b_g$ for all $g \in G$	$u_g \leq 0$ for all $g \in G$
variables $x_i \geq 0$ for all $i \in I$	constraints $A_{.i}^T u \geq c_i$ for all $i \in I$
x_j free for all $j \in J$	$A_{.j} u = c_j$ for all $j \in J$
$x_k \leq 0$ for all $k \in K$	$A_{.k}^T u \leq c_k$ for all $k \in K$

(b) *If one of the programs $\mathcal{PP}$ or $\mathcal{DP}$ has no feasible solution, then the other is either unbounded or has no feasible solution.*

(c) *If one of the programs is unbounded, then the other has no feasible solution.*

This theorem is equivalent to the famous **Farkas' lemma**:

Theorem 0.5.2 ([17]). *There exists a vector $x \in \mathbb{R}^n$ such that $Ax \leq b, x \geq 0$ if and only if there does not exist a vector $y \in R^m$ such that $y^T A = \mathbf{0}, y \geq \mathbf{0}$ and $y^T b < 0$.*

0.6 Solution Methods for Linear Programs

Let P be the polyhedron corresponding to $Ax \leq b$, i. e., $P = \{x \in \mathbb{R}^n \mid Ax \leq b\}$ is the geometrical interpretation of the constraint system $Ax \leq b$. Considering the set of optimal solutions of LP we see that it is a face of P. If P is nonempty and the LP bounded, there exists a vertex of P that is an optimal solution of the LP. The important question is: how to find this vertex?

The first documented linear programming problem was solved in 1947 when G. Dantzig formulated U. S. Air Force planning problems and designed the **simplex algorithm** for solving these problems. Soon it was discovered that this tool is convenient for a huge number of practical problems in various fields. For over 30 years it remained an open question whether linear programming problems are solvable in polynomial time or not. In 1979 L. G. Khachian modified an originally non-linear non-differential optimization method, known as the **ellipsoid method**, to prove the feasibility of a linear programming problem in polynomial time [21]. This was a great step but unfortunately the result was of no help in practice. Only a few years later in 1984 N. Karmarkar invented the **interior point method** [22] which is much faster than the ellipsoid method and in some cases even faster than the simplex method. Until today these three methods are the fundamental solution techniques for linear programming problems. For further literature we refer to [23–25]. The most frequently used is the simplex algorithm which will be described now.

0.6.1 Simplex Algorithm

The basic idea is to start from a vertex of the corresponding polyhedron and jump to a neighboring vertex with a better objective function value until we reach an optimum. This "jump" is realized by replacing one index of the basis B. To define the basis of P let T be the index set of the columns of A. Further let m be the rank of A. A **basis of a linear program** is a subset $B \subseteq T$ with $|B| = m$. The variables belonging to B are called **active variables**; all others are **inactive variables**. Now denote the columns of A corresponding to B by A_B. For a set with only one element $B = \{v\}$, we write A_v instead of $A_{\{v\}}$. Analogously define c_B, x_B, and b_B. Fixing $x_i = 0$ for all $i \in T \setminus B$ and determining the unique solution of $A_B x_B = b_B$ we obtain a solution of $Ax = b$ which is known as the **basis solution of** B. A basis is a **feasible basis** if the basis solution is a feasible solution of the LP.

The simplex algorithm generates a sequence of feasible bases and tests in each step whether the corresponding dual solution is feasible for $\mathcal{DP}$ or not. If it is, the current basis solution is optimal, due to Theorem 0.5.1-(b). Suppose we have a basis and a feasible basis solution, but it is not optimal for our minimization problem. We first choose an index $i \in T \setminus B$ with $y^T A_i > c_i$ and save it to enter the basis B. Now we have to determine an index j leaving the current basis. This procedure is sometimes called the **pricing step** or **ratio test**. We start by finding the unique solution to $A_B z = A_i$, i.e., we want to know how the new column A_i can be described in terms of the current basis. Then, we choose a real value $\varepsilon \in \mathbb{R}$ and replace the current basis solution x_B by $x_B - \varepsilon z$ and $x_i = \varepsilon$. This leads to the following changed objective function value

$$c_B^T(x_B - \varepsilon z) + c_i \varepsilon = c_B^T x_B + \varepsilon(c_i - c_B^T z)$$
$$= c_B^T x_B + \varepsilon(c_i - y^T A_B z)$$
$$= c_B^T x_B + \varepsilon(c_i - y^T A_i).$$

By the choice of i, we have $c_i - y^T A_i < 0$ and hence every positive ε decreases, i.e., improves, the objective function value. As we want to take the best possible ε, we search for the largest ε such that $x_B - \varepsilon z \geq 0$.

If no such ε exists, the LP is unbounded. If we can find such an ε, there exists an index $j \in B$ with $z_j > 0$ and $(x_B - \varepsilon z)_j = 0$. We replace this index j with the chosen index i and repeat the procedure. Algorithm 1 (taken from [18]) shows a summary of the simplex algorithm.

If the algorithm terminates, it obviously returns the optimal solution or the unboundedness of the LP. If we determine $\varepsilon = 0$ at any iteration step of the algorithm, we obtain the same basis we already have and the simplex algorithm might not terminate. To prevent this cycling, certain pivot rules have to be applied. Two things of the simplex method are essential for its success: choose a feasible initial basis and improve the objective function value in each step.

The above described simplex algorithm is commonly known as **phase II** of the **primal simplex algorithm**. That is because we skipped phase I by assuming that a feasible basis solution exists. Furthermore the algorithm starts with primal feasibility and stops immediately if dual feasibility has been ensured. The reverse procedure is the so-called **dual simplex algorithm**.

There exist several implementations of the simplex algorithm. A powerful and widely used software is the commercial CPLEX code from ILOG. It cannot only be used for solving linear programs but also for solving mixed integer, quadratic and mixed integer quadratic programs [26]. A non-commercial implementation of the simplex algorithm is SoPlex (The **S**equential **o**bject-oriented sim**plex** class library). "SoPlex is an implementation of the revised simplex algorithm. It features primal and dual solving routines for linear programs and is implemented as a C++ class library that can be used with other programs."[27, 28]

0.6.2 Pricing

In many problems the corresponding LP consists of too many variables, i.e., they cannot all be written down explicitly or even generated. The key idea is now to generate variables on demand. The reason behind this procedure is that

---
Algorithm 1 `Simplex Algorithm with an Initial Basis`

Input: Current LP solution x^* for the minimization problem,
Feasible basis B,
Output: Optimal solution of the linear program if it exists.

1: **loop**
2: Find the unique solution to $y^T A_B = c_B$.
3: **if** $y^T A_i \geq c_i$ for all i not in B **then**
4:   **return** Optimal solution x^*. // Current solution x^* is optimal.
5: **else**
6:   Choose $i \notin B$ such that $y^T A_i < c_i$.
7:   Find the unique solution to $A_B z = A_i$.
8:   Find the largest ε such that $x_B - \varepsilon z \geq 0$.
9:   **if** ε does not exist **then**
10:    **return** The LP is unbounded.
11:   **else**
12:    Choose $j \in B$ such that $z_j > 0$ and the jth component of $x_B - \varepsilon z$ is 0.
13:    Replace B by $(B \cup \{i\}) \setminus \{j\}$ and x_B by $x_B - \varepsilon z$ and $x_i = \varepsilon$.
14:   **end if**
15: **end if**
16: **end loop**
---

most of the variables are non-basis variables and equal to 0 in the optimal solution. Therefore one considers only a subset of variables, to be more precise: one generates only those variables that have the potential to improve the objective function value. The knowledge whether a variable can enhance the objective can be obtained from the simplex algorithm.

We define the **reduced cost of a variable** by $\mu_i := c_i - y^T A_i$. The change of the objective function within an iteration step of the simplex algorithm is $c_B^T x_B + \mu_i$. If μ_i is non-negative for all indices $i \in T \setminus B$, the objective function value cannot be decreased, i.e., no improvement is possible. Hence the current basis solution is optimal and we say that all non-basis variables **price out correctly**.

The so-called **pricing problem** consists in the test whether all variables price out correctly, or in determining a non-basis variable that does not price out correctly. Pricing means to start the simplex algorithm with a small set of variables and **price-in** all necessary variables to prove optimality.

0.6.3 Pricing Versus Column Generation

The important difference of pricing and column generation is that in the first case the whole coefficient matrix of the LP is known, whereas this is not the case if column generation is done. To be more precise, pricing denotes the process of determining the missing columns. **Column generation** denotes the procedure of determining the missing columns and their unknown coefficients. This makes a huge difference. In the case of pricing one "only" has to consider the reduced costs of all non-basis variables. In column generation one has to formulate a so-called restricted master (or primary) problem whose solution yields the dual multipliers needed in a current pricing (or secondary) problem. As described above the pricing problem determines the missing columns and the expanded restricted master problem can then be solved again. Possible procedures are Dantzig–Wolfe decomposition [29] or Benders decomposition [30]. Unfortunately the implementations are difficult and the convergence of the algorithms is often slow. Nevertheless solution strategies for huge, partly unknown LPs exist.

0.7 Integer Programming

There is only a small but significant difference between a linear programming problem and an integer programming problem. An **integer programming problem**, for short **integer program**, consists of finding a vector $x \in \mathbb{Z}^n$ that fulfills all given constraints $Ax \leq b$ and maximizes a certain objective function $c^T x$, i. e.,

$$(\mathcal{IP}) \quad \begin{aligned} \max \quad & c^T x \\ \text{s.t.} \quad & Ax \leq b \\ & x \in \mathbb{Z}^n, \end{aligned}$$

where the last condition is the **integrality constraint**. Unfortunately this integrality constraint makes the problem much harder than its linear programming variant. To be more precise: in general integer programs are $\mathcal{NP}$-hard [31]. This is true although the number of feasible solutions is, for a bounded polyhedron, finite — in contrast to the infinite number of feasible solutions for a linear program.

As soon as linear programming was invented, integer programming became an important topic. This is due to the fact that there exists a huge number of practical applications for integer programming. The most famous is without question the Traveling Salesman problem (see [32, 33]) but from production planning to timetable scheduling one can find countless examples that are relevant today. Therefore a lot of effort has been and is still made to develop solution strategies for integer programs.

0.8 Solution Methods for Integer Programs

Solution methods for integer programs can be partitioned into exact and heuristic algorithms. Heuristic algorithms compute relatively "good" results within a short running time. Unfortunately a heuristic itself cannot evaluate its solution, i. e., we do not know whether a produced solution is near to or far from the optimal solution. Nevertheless heuristics are important to accelerate exact algorithms. Moreover in several applications a short running time is more important than an exact solution. Therefore a well implemented heuristic can surely be a powerful tool. The following presentation of exact algorithms is based on [34, 35].

0.8.1 Branch-and-Bound

The classical exact algorithm for solving integer programming problems is the so-called **branch-and-bound** algorithm, often known as **explicit enumeration**. It is based on the ancient idea to solve a problem using the **divide-and-conquer** principle. Although A. H. Land and A. G. Doig redeveloped the idea [36], whereas the first branch-and-bound algorithm was formulated by R. J. Dakin in 1965 [37].

The idea is to split the original problem into smaller **subproblems** by successively fixing the variables to integer values. Each fixing leads to two new subproblems that have to be considered. In this way one obtains the so-called **branch-and-bound tree**. Solving the corresponding LP of the minimization problem at a certain node in this tree leads to a **local lower bound**. Now we compare this local lower bound with the **global upper bound**, i. e., with a feasible solution obtained by a heuristic for the original problem. If the local lower bound is greater than the global upper bound, we know that the optimal solution cannot be within the node's branch-and-bound subtree. In this case we **fathom** the current node, i. e., we do not consider it and its subtree anymore. This saves a lot of running time if the upper bound is strong. Figure 1 displays an overview of the branch-and-bound algorithm for a minimization problem.[1]

We say a **variable is fixed** if it has this value for the rest of the optimization. If this holds in the current branch-and-bound node and in its subtree only, we say the **variable is set**.

---

[1]This figure has been adopted from [35].

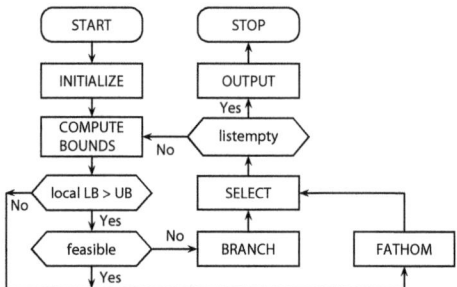

Figure 1: Flowchart of a branch-and-bound algorithm.

There exist several different branching schemes. The simplest is to choose any fractional variable x_i^* of the current solution and define the two subproblems in each branching step in the following way. The first subproblem consists of the constraint system with the new constraint $x_i \geq \lceil x_i^* \rceil$. The second subproblem is constructed adding the constraint $x_i \leq \lfloor x_i^* \rfloor$ to the LP-relaxation. This general procedure is called **branching on a variable** and leads to a binary branch-and-bound tree.

Instead of branching on a variable one can **branch on constraints** as well. In this case each subproblem is defined by one additional constraint, where an arbitrary constraint set can be chosen as long as it fulfills the general branching principle. This means that the (not necessarily disjoint) union of the feasible solutions of all subproblems equals the set of all feasible solutions in the branch-and-bound node.

The choice of the branching variable or the branching constraint is of great importance. Unfortunately none is useful for all kinds of problems. Problem specific selection strategies can be found in Section 4.5.4.

Unfortunately the number of subproblems in a branch-and-bound tree can be very large. Furthermore, the algorithm is strongly dependent on good bounds. Therefore a branch-and-bound algorithm is often expensive concerning memory usage and running time. Nevertheless it is an intuitive approach for solving combinatorial optimization problems. An important advantage is that a guarantee for the current best solution is given at any time.

0.8.2 Cutting Planes

A second and equally powerful exact algorithm for solving integer programs is the **cutting plane method**. Its central idea is the so-called relaxation. A **relaxation** of an integer or linear program means that some constraints of the system are missing. The most famous relaxation is the so-called **LP-relaxation** in which the integrality constraints are deleted from the integer program. Its importance is due to the fact that we can use all solution methods of the linear programming theory to solve the LP-relaxation. Note: for a minimization problem, the optimal solution of an LP-relaxation cannot be larger than any feasible solution of the original program.

Consider an optimal solution of an LP-relaxation that is fractional. We now search for a valid constraint of our original problem that cuts off the optimal solution of the LP-relaxation. This identification of a violated (with respect to the solution of the LP-relaxation) and valid (for the original problem) constraint is called **separation problem**. The identified constraint is called a **cut** or cutting plane. Having found one or several cuts, we add them to the LP-relaxation, re-optimize and separate again. Unfortunately the complexity of the separation problem is equivalent to the complexity of solving the original problem to optimality. Furthermore the algorithm is strongly dependent on the strength of the cutting planes. Another drawback is that one has to be careful to avoid rounding errors and numerical problems.

We will now consider the combination of the algorithms described above.

0.8.3 Branch-and-Cut

The so-called **branch-and-cut method** is the most important and surely most applied algorithm for solving integer programs to optimality. Being a combination of branch-and-bound and cutting plane method it inherits the properties of its parts, i.e., its running time depends on good bounds and strong cuts. The two algorithms support one another in the following way. The branch-and-bound part determines those subtrees that are worth searching for the optimal solution. The cutting planes improve the local lower bounds remarkably and accelerate therefore the search within the subtrees significantly.

In practice one separates cutting planes in the current node and branches if the LP solution is still fractional until no violated cuts can be found anymore. To tackle the phenomenon that several cutting planes are added but the objective function does not improve significantly, we can make use of **tailing-off**. This means if for the last k iterations the objective function value improved less than l%, the separation of cutting planes is stopped and branching is performed instead. The choice of the parameters k and l is problem specific. In combination with a good order in which the different constraint types are separated and a smart choice of the number of separated cuts per iteration, one can shorten the running time of the algorithm by about 30%.

The **gap closure** of a minimization problem is the percentage by which the gap between the upper bound c_{best} and an old lower bound c as well as a new lower bound c' could be closed:

$$\text{gap closure} = 100\% - \frac{|c' - c_{\text{best}}|}{|c - c_{\text{best}}|} \times 100\%.$$

It is obvious that the separation increases the computational effort per node. On the other hand the cutting planes help to improve the LP bound and therefore reduce the number of nodes that have to be considered. Therefore a branch-and-cut algorithm is in general more efficient than a branch-and-bound or a cutting plane algorithm. In many cases the largest part of the computation is to prove the optimality of an early found optimal solution.

Within this thesis the branch-and-cut framework ABACUS 2.4 (**A B**ranch-**A**nd-**CU**t **S**ystem) [35, 38] is used in combination with the LP solver CPLEX 8.1 [26].

0.8.4 Branch-and-Cut-and-Price

The **branch-and-cut-and-price** algorithm combines the advantages of a branch-and-cut algorithm with the possibility to work with huge LPs. Unfortunately this name has been widely used for branch-and-cut algorithms combined with column generation. Therefore recall the differences between column generation and pricing described in Section 0.6.3. Within this thesis we will not consider column generation as the whole coefficient matrix of the LP is known.

The branch-and-cut-and-price algorithm is especially useful for a certain class of combinatorial optimization problems. These are problems with a large number of variables and sparse feasible solutions. A prominent example is the Symmetric Traveling Salesman Problem. It has $\binom{n}{2}$ variables but in a tour only n of them are non-zero. Here a suitable subset of start variables is chosen and the algorithm prices in more variables if it is required for the correctness of the algorithm. This method is often denoted as **sparse graph technique**. See [39] for a corresponding procedure for the Travelings Salesman problem.

The work flow of a branch-and-cut-and-price algorithm for the use of sparse graph techniques is presented in Figure 2. It is based on the corresponding figure in [35], where a detailed description of the single parts can be found. Note: pricing is necessary before a node can be fathomed. Additional pricing steps can be performed, which might improve the performance of the algorithm. This was the case for the Traveling Salesman problem for example [39]. The impact of additional pricing steps for the Minimum Linear Arrangement problem is investigated in Section 5.5.6.

Branch-and-Price

When combining branching and pricing, one has to take care of some specialties. The first is based on the fact that the pricing problem has to be solved not only in the root but in every node of the branch-and-cut tree. As a consequence

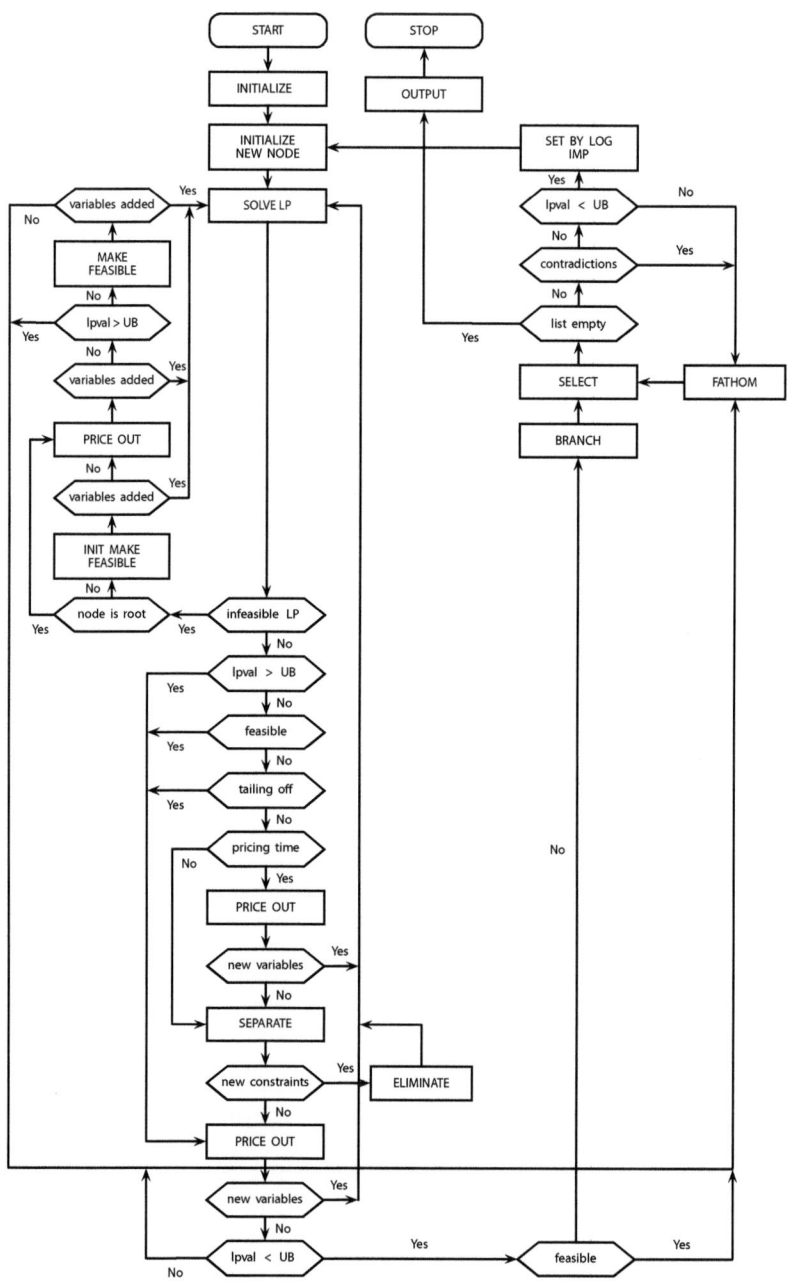

Figure 2: Flowchart of a branch-and-cut-and-price algorithm.

0.8. Solution Methods for Integer Programs

a variable x_i might be set to 0 in this subtree although it has negative reduced cost in the current node's LP solution. It should therefore be added to the LP, which is not possible within this subtree. To tackle this case we have to be careful during the pricing step: If a variable should be added due to its reduced costs, we have to check whether it is fixed or set. Only if it is not fixed or set, we add it to the LP. Alternative branching schemes that are compatible with pricing can be found in [40–42]. The second challenge is that the LP can be infeasible due to fixed or set variables. How we can cope with this infeasibility is described in the following.

Infeasible Linear Programs

There are two reasons for infeasibility. The first is that the left hand side of a violated "≥"-constraint might be empty. This can occur when variables are fixed and set during the branching phase. Another reason for infeasibility is that the LP solver finds a dual feasible basis that is primaly infeasible. These *false* infeasibilities can be resolved by activating additional variables. The important question is how to find the right variables? An overview of the general handling of infeasibility is given in Algorithm 2 InfeasibleLP().

Algorithm 2 InfeasibleLP()

Input: Constraint system (A,b) with infeasible subsystem $(A_{\frac{i}{2}}, b_{\frac{i}{2}})$,
  Objective function min $c^T x$, where $c \geq 0$,
  Global upper bound gUB,
  Current primal variables x^*,
  Current dual variables y^*,
  Current dual feasible basis B.
Output: New constraint system (A', b') if feasibility can be restored by new variables,
  False otherwise.

1: **if** #subproblems > 1 and InitMakeFeas($A_{\frac{i}{2}}, b_{\frac{i}{2}}, c$) == True (see Alg. 3) **then**
2:   **return** New constraint system (A', b'). // Variables were added.
3: **else**
4:   **if** pricing() == True **then**
5:     **return** New constraint system (A', b'). // Variables were priced in.
6:   **else**
7:     **if** $c^T x^* \geq$ gUB **then**
8:       **return** False. // It is impossible to reestablish feasibility.
9:     **else**
10:      **if** MakeFeasible(A,b,c,x^*,y^*,B) == True (see Alg. 4) **then**
11:        **return** New constraint system (A', b'). // Variables were added.
12:      **else**
13:        **return** False. // It is impossible to restore feasibility.
14:      **end if**
15:    **end if**
16: **end if**
17: **end if**

If the infeasibility is due to the fixing and setting of all involved variables, we apply the procedure InitMakeFeas() displayed in Algorithm 3. Let x^* be the current LP solution. Suppose $ax \leq \beta$ is an infeasible constraint with void left hand side. That means all coefficients of a corresponding to active variables are 0. Now all variables not part of the constraint system are scanned whether their coefficient in $ax \leq \beta$ is negative. If this is the case this variable might restore the feasibility and is therefore added to the LP. To make the best use of InitMakeFeas() we formulate as many constraints as possible in the form $ax \leq \beta$ with a having negative coefficients.

Algorithm 3 `InitMakeFeas`$(A_{\sharp}, b_{\sharp}, c)$

Input: Constraint system (A,b) with infeasible subsystem $(A_{\sharp}, b_{\sharp})$,
Objective function coefficients c.
Output: True if feasibility might be restored by new variables,
False otherwise.

1: **for all** rows $ax \leq \beta$ of $(A_{\sharp}, b_{\sharp})$ **do**
2:    **for all** inactive variables v **do**
3:      **if** $a_v < 0$ **then**
4:        Add variable v.
5:        Go to step 2. // Variable found which might restore feasibility of $ax \leq \beta$.
6:      **end if**
7:    **end for**
8:    **return** False. // No variable can restore feasibility of $ax \leq \beta$.
9: **end for**
10: **return** True.

If the infeasibility has another reason, we solve the pricing problem and add all variables to the LP that have negative reduced costs. If this does not help, we have to check whether the current LP solution x^* satisfies the bounding condition, i.e., $c^T x^* \leq$ gUB. If not, we continue by considering the **infeasible variable** with the goal to make it feasible again. A variable is called infeasible, if its value is less than its lower bound or greater than its upper bound. The goal is to change the LP in such a way that every infeasible variable is feasible again. How we try to achieve this is summarized in Algorithm 4. The idea is to add a new variable v for every infeasible variable $v_{\sharp}$, whereas the new objective function value must not be too high, i. e., $c^T x^* + r_v \leq$ gUB. The additional variable v has the property that it can adjust the infeasibility of the infeasible variable $v_{\sharp}$. If the infeasible variable's value is too low, the component $(A_B^{-1} A_v)_{v_{\sharp}}$ of the new variable v has to be greater than zero (recall the notations in Section 0.6.1). Otherwise the new variable cannot compensate for the infeasibility of $v_{\sharp}$. If the value of the infeasible variable is too high, the new variable's contribution has to be negative to be able to restore feasibility. Unfortunately we cannot assure that other components of the new variable do not destroy the helpful impact of the v^*th component. Therefore condition (8) of Algorithm 4 is a necessary but not sufficient condition for the reduction of the infeasibility. Hence this procedure is only heuristic and not exact. Note the new variable v is not determined by its reduced cost but by simulating one iteration of the dual simplex algorithm.

If for an infeasible variable $v_{\sharp}$ no new variable could be found, the procedure can be stopped as the LP is then indeed infeasible.

0.8. Solution Methods for Integer Programs

Algorithm 4 `MakeFeasible`(A,b,c,x^*,y^*,B)

Input: Infeasible constraint system (A,b),
Objective function coefficients c,
Current primal variables x^*,
Current dual variables y^*,
Current dual feasible basis B.

Output: True if feasibility might be restored by new variables,
False otherwise.

1: Let A_B be the columns of A corresponding to variables in B.
2: Let c_v be the objective function coefficient of v.
3: Let $\mu_v := c_v - (y^*)^T A_v$ be the reduced cost of variable v.
4: Let LB_v be the lower bound of variable v.
5: Let v_i be one of the infeasible variables.
6: **for all** inactive variables v **do**
7:   **if** $c^T x^* + \mu_v \leq \text{gUB}$ **then**
8:     **if** $\left(v_i < \text{LB}_{v_i} \text{ and } (A_B^{-1} A_v)_{v_i} > 0\right)$ or $\left(v_i > \text{UB}_{v_i} \text{ and } (A_B^{-1} A_v)_{v_i} < 0\right)$ **then**
9:       Add variable v.
10:       **return** True. // Variable might restore feasibility.
11:     **end if**
12:   **else**
13:     Do not add variable v. // New LP value would be too high.
14:   **end if**
15: **end for**
16: **return** False. // It is impossible to reestablish feasibility.

Chapter 1

A Brief Survey

1.1 Definition of MinLA and Basic Properties

Let $G = (V,E)$ be an undirected, weighted graph on n nodes with m edges and non-negative edge weights c_{ij} for $ij \in E$. The Minimum Linear Arrangement problem consists in finding a linear ordering of the nodes of a given graph such that the sum of the weighted edge lengths is minimized. To be more precise, the goal is to find a one-to-one mapping $\pi : V \to \{1,\ldots,n\}$ that maps the nodes of the graph to the set $\{1,\ldots,n\}$ and minimizes the sum over all weighted edge lengths concerning π:

$$\min_{\pi \in S(n)} \sum_{ij \in E} c_{ij} |\pi(i) - \pi(j)|. \tag{1.1}$$

The mapping π is called an **arrangement**, **layout** or **labeling**. The problem of determining an optimal labeling is called the **Minimum Linear Arrangement problem**, short **MinLA**, and was introduced by Harper [43]. In the field of approximation algorithms, the abbreviations **MLA** and **OLA** are widely used. Furthermore, the following synonyms are established: **Minimum Length Linear Arrangement problem**, **Minimum Length Layout problem**, **Total Edge Length problem**, **Edge Sum problem**, **Minimum 1-Sum problem**, **Dilation Minimization problem**, **Graph Ordering problem** and **(Optimal) Linear Ordering**. Figure 1.1 shows an example of a graph and its minimum linear arrangement solution.

The optimal value of the objective function (1.1) for the input graph G is denoted by $c_{\mathrm{opt}}(G)$. We distinguish between the weighted version of this optimization problem and its unweighted version, in which all coefficients c_{ij} are equal to 1. If the input graph G is obvious we omit its declaration. If G is not connected the MinLA can be solved independently for every component. The optimal objective function value $c_{\mathrm{opt}}(G)$ is the sum of the optimal objective function values of its components. We therefore assume all input graphs to be connected.

The Linear Arrangement problem can be formulated as special case of the General Placement problem [44], the Quadratic Assignment problem [44] and the Single Row Facility Layout problem [45]. It can be generalized to the Storage Time Product Minimization problem [46]. The MinLA is one of the most important graph layout problems.

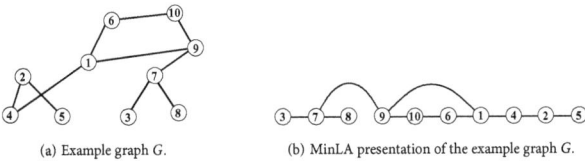

(a) Example graph G. (b) MinLA presentation of the example graph G.

Figure 1.1: An example graph and its MinLA presentation.

Its most famous variant is the **Matrix Bandwidth problem**, originally posed in [43], in which the maximal length of an edge is to be minimized

$$\min_{\pi \in S(n)} \max_{ij \in E} |\pi(i) - \pi(j)|.$$

Another interesting variant is its generalization into two dimensions. Hansen introduced this **Grid Arrangement problem** in 1989. In [47] a survey about this problem can be found.

Considering the reverse objective function of (1.1) leads to the **Maximum Linear Arrangement problem** which was investigated in [48]. The parameterized version of the MinLA is considered in [49–52]. Here, for a given graph and a non-negative integer number k the goal is to answer the question, whether there exists a linear arrangement of G with objective function value less or equal than k, i.e.,

$$\text{does } \pi \in S(n) \text{ exist such that } \sum_{ij \in E} c_{ij} |\pi(i) - \pi(j)| \le k?$$

1.2 Applications

Harper defined the problem in 1964 to develop error-correcting codes with minimal average absolute errors. Ten years later it became important for the VLSI technology, as it was considered as a simplified mathematical model of the placement phase, where the nodes represent the modules and edges of the graph correspond to the interconnections [53]. Today's main application of the MinLA is in the area of graph drawing, as the sum of all edge lengths can be seen as a criterion for a good presentation of a graph. It finds other applications in a range of fields, including the layout of UML sequence diagrams, software diagram layout in general and especially for entity relationship models [54] and data flow diagrams [55]. Certain tasks of wiring problems and within communication systems can be solved with the MinLA as well. It has been shown to be relevant to solve the Single Machine Job Scheduling problem [46, 56]. Furthermore, it has even been used in computational biology [57], for example as an over-simplified model of some nervous activity in the cortex [58].

1.3 Complexity

The MinLA is a classical $\mathcal{NP}$-hard optimization problem. Its decision problem is $\mathcal{NP}$-complete [59]. Compared to other $\mathcal{NP}$-hard optimization problems the MinLA turns out to be more difficult and is extremely hard to solve in practice. The fastest exact algorithm is based on dynamic programming and has a running time of $O(2^n m)$ [3].

1.3.1 $\mathcal{NP}$-hard Cases

The complexity of the problem remains the same even for bipartite graphs [60]. But there are special cases in which the MinLA can be solved efficiently. We present several of these in the following.

1.3.2 Polynomially Solvable Cases

As every edge has at least length 1 and at most length $n-1$ within a linear arrangement we obtain the lower and upper bound $m \le c_{\text{opt}} \le m(n-1)$.

For the **complete graph** K_n all $n!$ arrangements are optimal with $c_{\text{opt}}(K_n) = (n+1)n(n-1)/6$. Given a **path** P^n with n nodes the optimal linear arrangement is the identity of the path, hence $c_{\text{opt}}(P^n) = n-1$ as all edges have length 1.

If one considers the kth power of the path graph with n nodes $(P^n)^k$, it was proven by Juvan & Mohar [61] that $c_{\text{opt}}((P^n)^k) = k(k+1)(3n-2k-1)/6$. Considering a **cycle** C^n there exist n optimal linear arrangements which have exactly one edge of length $n-1$ and $n-1$ edges of length 1, see Figure 1.2. Therefore, $c_{\text{opt}}(C^n) = 2(n-1)$. A **star** S^n has one center node of degree $n-1$ and $n-1$ nodes that are adjacent to the center, i.e., $S^n = K_{1,n-1}$ is a special bipartite

1.3. COMPLEXITY

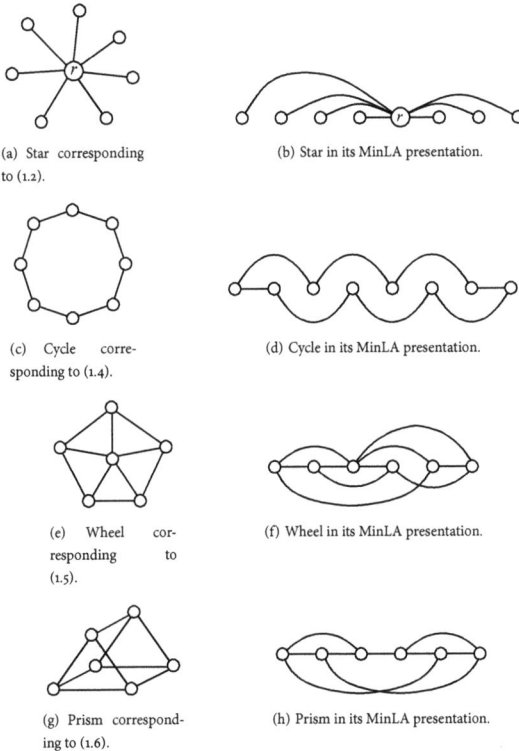

(a) Star corresponding to (1.2).

(b) Star in its MinLA presentation.

(c) Cycle corresponding to (1.4).

(d) Cycle in its MinLA presentation.

(e) Wheel corresponding to (1.5).

(f) Wheel in its MinLA presentation.

(g) Prism corresponding to (1.6).

(h) Prism in its MinLA presentation.

Figure 1.2: Several graphs and their MinLA presentation.

graph. The best possible arrangement is therefore to have two edges of length 1, two of length 2, two of length 3 and so forth. This leads to the objective function value $c_{\text{opt}}(S^n) = \lfloor n^2/4 \rfloor$.

A k-**wheel** W^k is a graph of $n+1$ nodes consisting of a star, and a cycle that contains all nodes but the center node. Liu & Vannelli [6] showed $c_{\text{opt}}(W^k) = c_{\text{opt}}(S^k) + c_{\text{opt}}(C^k) + 1$. A k-**prism** P^k_{rism} is defined as the union of two k-edge cycles $C_1 = u_1, \ldots, u_{k+1}$ and $C_s = v_1, \ldots, v_{k+1}$, where $u_{k+1} = u_1$ and $v_{k+1} = v_1$, and all edges $\{u_i v_i \mid i = 1, \ldots, k\}$. In [6], it is proven that

$$c_{\text{opt}}(P^k_{\text{rism}}) = \begin{cases} 17, & k = 3, \\ 6 + 4(k-1) + 5(k-2), & k \geq 4. \end{cases}$$

Although the MinLA is $\mathcal{NP}$-hard for bipartite graphs $K_{m,n}$ with arbitrary n and m, the case $m \geq n$ is an exception. In [61], it is proven that

$$c_{\text{opt}}(K_{m,n}) = \begin{cases} n(3m^2 + 6mn - n^2 + 4)/12, & m+n \text{ even,} \\ n(3m^2 + 6mn - n^2 + 1)/12, & m+n \text{ odd.} \end{cases}$$

Table 1.1 on page 22 summarizes several more instances for which the MinLA can be solved to optimality in polynomial time.

Table 1.1: Time complexity of algorithms that solve the MinLA of certain graphs to optimality. (With δ being the number of biconnected components containing a central articulation point.)

Graph	Complexity	Reference
Hypercubes	$O(n)$	[62]
Square meshes	$O(n)$	[58]
Rectangular meshes	$O(n)$	[63]
Trees	$O(n^{\log 3/\log 2})$	[64]
Rooted trees	$O(n \log n)$	[65]
Complete k-level 3-ary trees	$O(n)$	[66]
d-dimensional c-ary clique	$O(n)$	[67]
DeBruijn graphs of degree four	$O(n)$	[68]
Certain Halin graphs	$O(n^2)$	[69]
Outer planar graphs	$O(\delta^2 n + n^2)$	[70]
Proper interval graph	$O(n)$	[71]

1.4 Computation of Upper Bounds

The practical significance of the MinLA has been a great motivation to address the problem in reasonable time. Therefore a lot of effort has been made to achieve near-optimal solutions.

1.4.1 Heuristics

We distinguish between heuristics, applicable to all sorts of problems and those which were developed particularly for the MinLA. The first are called **meta heuristics**, which include the well-known **local search heuristics**. Table 1.2 summarizes the work that has been done with this sort of heuristics. Several **MinLA specific heuristics** are presented in Table 1.3. Some of the presented heuristics perform very well in practice, see [73, 83, 84] for good surveys.

1.4.2 Approximations

It is not known, whether the MinLA can be approximated within a constant factor in polynomial time. In [85] it is conjectured that it is not true. It was proven in [86] that a PTAS for dense graphs exists. To be more precise: An approximation within a $1 + \varepsilon$ factor can be computed in time $n^{O(1/\varepsilon)}$ for any $\varepsilon > 0$. Unfortunately, this result cannot be generalized. It was shown in [87] that no PTAS exists for an arbitrary input graph with the standard assumption that $\mathcal{NP}$-complete problems cannot be solved in randomized sub-exponential time. In [88] an algorithm based on spreading metrics with an approximation ration of $O(\log n)$ is presented. The same ratio was obtained by [89] using flow metrics. If the graph is restricted to be planar, an $O(\log \log n)$ approximation algorithm can be found [88]. A combination of the techniques in [90] with the rounding algorithms of [88] leads to the currently best approximation $O(\sqrt{\log n} \log \log n)$ [91]. In [92] the same upper bound was obtained independently. A good overview over approximation algorithms for the MinLA can be found in [83].

The integrality gap of a linear program is the worst-case ratio between the optimum of the integer program and a relaxation of this formulation. The integrality gap $\Omega(\log \log n)$ [93] for the MinLA was obtained with a semi definite programming relaxation, see [91, 92] for more details.

Table 1.2: Overview of different meta heuristics for the MinLA.

Type	Name	Abbrev.	Reference
-	Successive augmentation	SCA	[2]
-	Spectral sequencing	SS	[61, 72]
Local search	Simulated annealing	SA	[2, 72, 73]
Local search	Genetic algorithms	GA	[1, 74]
Local search	Hill climbing	HC	[1, 73]
Combination	SS+SA		[73, 75]
Combination	GA+HC		[74]
Combination	GA+Dynamic programming		[76]

Table 1.3: Overview of different MinLA specific heuristics.

Name	Abbrev.	Reference
Decomposition tree	DT	[77, 78]
Multi scale/Multi level	MS	[3, 4]
Algebraic multi grid scheme	MG	[4]
Memetic algorithms	MA	[79]
MA with different recombination operators	TX	[80]
Frontal increase minimization	FIM	[81, 82]

1.5 Computation of Lower Bounds

The reason for the big gap for unsolved instances is probably the lack of good, i. e., strong and efficiently computable lower bounds. In this section we present the efforts that have been made to develop methods to obtain such lower bounds.

1.5.1 Combinatorial Bounds

Degree Lower Bound

The degree lower bound was first formulated in [64]. Its key idea is the fact that the degree of each node leads to a lower bound of the distances at each node. If, for example, node i has degree $\deg(i) = 5$, the sum of edge lengths of those edges starting at node i is at least $(2 \times 1) + (2 \times 2) + (1 \times 3) = 9$. This is due to the fact that we cannot have more than two edges of each length starting at the same node. The general formula for a node i is $\lfloor (\deg(i) + 1)^2 / 4 \rfloor$. We can now sum up all lower bounds of nodes and obtain a lower bound of the MinLA. As every edge has two adjacent nodes, every edge appears in two such inequalities. We therefore have to divide the whole sum by two:

$$\frac{1}{2} \sum_{i \in V} \lfloor (\deg(i) + 1)^2 / 4 \rfloor.$$

Edge Lower Bound

This straightforward lower bound is considered in [73] and called edge lower bound here. It is based on the observation that for a graph with n nodes and m edges there exist at most $n - 1$ edges of length one, at most $n - 2$ of length two and so forth. Summing up these edge lengths one obtains a lower bound for the MinLA. E. g. for $n = 8$ and $m = 17$ we have at most seven edges of length 1, six of length 2 and four of length 3, i. e., $(7 \times 1) + (6 \times 2) + (4 \times 3) = 31$ is a lower bound for the objective function value. This bound can easily be computed.

Gomory-Hu-Tree Lower Bound

This bound is often called **cut tree bound** and is based on the computation of a Gomory-Hu-tree [94]. It was first described in [65]. The key idea is that the sum of all weights in the Gomory-Hu-tree is a lower bound for the MinLA. A detailed presentation of this bound can be found in [6]. In the same paper a variant of this type of lower bound is constructed. It is based on minimum cuts in the graph and is tighter than the Gomory-Hu-tree bound.

Path Lower Bound

We have seen in Section 1.3.2 that the lower bound of the linear arrangement for the kth power of a path graph is proven to be $c_{\text{opt}}((P^n)^k) = k(k+1)(3n - 2k - 1)/6$. Now let $k_{\max}$ be the largest k such that $|E((P^n)^k)| \leq m$, where $m = |E(G)|$. [61] proved a theorem that says $c_{\text{opt}}(G) \geq c_{\text{opt}}((P^n)^k)$ for $k = \lfloor k_{\max} \rfloor$.

Mesh Lower Bound Method

We have seen in Table 1.1 that the MinLA can be solved in polynomial time for meshes. Let M^n be a square mesh of side n, i. e., $V(M^n) = [n]^2$ and $E(M^n) = \{ij \mid \|i - j\|_2 = 1\}$. It has been shown in [63] and [58] that $c_{\text{opt}}(M^n) = (4 - \sqrt{2})n^3/3 + O(n^2)$. The idea is now to decompose the original graph G into k disjoint square meshes $M_1, \ldots, M_k$ as $c_{\text{opt}}(G) \geq \sum_{i=1}^{k} c_{\text{opt}}(M_i)$ and $c_{\text{opt}}(M_i)$ is known. A suggestion for a practical realization of this idea can be found in [73].

Unfortunately none of the presented bounds is generally the best. If one considers instances that have quite different structures the best results are obtained by different lower bound algorithms. The degree lower bound is, for

1.5.2 Linear Programming Bounds

We start the presentation of linear programming bounds by introducing the most common variables for modeling the MinLA problem.

Integral Distance Variables y

The integral distance variables are defined as

$$y_{ij} := |\pi(i) - \pi(j)| \quad \text{for all } i < j \in V.$$

They have been used in various formulations of the MinLA problem and will be part of the model presented in Chapter 2 as well. We will now state some inequalities that can easily be formulated with y-variables.

$$\sum_{rj \in E} y_{rj} \geq \left\lfloor \frac{|S|^2}{4} \right\rfloor \quad \text{for every star } (S,E), \tag{1.2}$$

where r is the center node of the star, see Figure 1.2.

$$\sum_{ij \in E} y_{ij} \geq \binom{|C|+1}{3} \quad \text{for every clique } (C,E), \tag{1.3}$$

$$\sum_{ij \in E} y_{ij} \geq 2|C'| - 2 \quad \text{for every cycle } (C',E), \tag{1.4}$$

$$\sum_{ij \in E} y_{ij} \geq \left\lfloor \frac{|W|^2}{4} \right\rfloor + 2|W| - 2 \quad \text{for every wheel } (W,E), \tag{1.5}$$

$$\sum_{ij \in E} y_{ij} \geq \begin{cases} 17, & \text{if } k = 3 \\ 9k - 8, & \text{if } k \geq 4 \end{cases} \quad \text{for every prism } (P,E), |P| = 2k. \tag{1.6}$$

The above stated constraints are called **star**, **clique**, **cycle** or **circuit**, **wheel** and **prism** inequality.

Besides these rank constraints so-called **hypermetric** and **bipartite** inequalities are known. The hypermetric inequalities are cut cone inequalities and have the original form $\sum_{i,j \in V} b_i b_j y_{ij} \leq 0$ where $\sum_{i \in V} b_i = 1$, see [95] for more information. They are called **pure** if $b_i \in \{0, \pm 1\}$ for all $i \in V$. Amaral & Letchford [45] the hypermetric inequalities are formulated for the Single Row Facility Layout problem, which is a generalization of the MinLA. For $S = \{i \in V \mid b_i = 1\}$ and $T = \{i \in V \mid b_i = -1\}$ one obtains

$$\sum_{i \in S, j \in T} y_{ij} - \sum_{ij \in E(T)} y_{ij} - \sum_{ij \in E(S)} y_{ij} \geq 0 \quad \text{for all } S,T \subset V, S \cap T = \emptyset, |T| = |S| - 1. \tag{1.7}$$

When $S = 2$ and $T = 1$, the pure hypermetric inequalities reduce to the well-known **triangle** inequalities

$$y_{ij} + y_{jl} \geq y_{il} \quad \text{for } i,j,l \in V.$$

Note the difference between the integral distance variables y and the ones used in [45]. The bipartite inequalities are similar to the hypermetric inequalities:

$$\sum_{i \in S, j \in T} y_{ij} - \sum_{ij \in E(S)} y_{ij} - \sum_{ij \in E(T)} y_{ij} \geq |S| \quad \text{for all } S,T \subset V, S \cap T = \emptyset, |T| = |S|. \tag{1.8}$$

Despite their name these inequalities are not constraints for bipartite subgraphs, see Figure 1.3, where the dotted lines have coefficient −1 and the others have coefficient +1. To obtain larger hypermetric inequalities we implement a heuristic suggested by Letchford [96], which originally was developed by Helmberg & Rendl [97]. We apply this

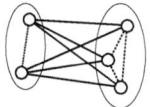

Figure 1.3: Example for the bipartite inequalities of the MinLA considered in this thesis.

idea to larger bipartite inequalities as well. In addition, we use the heuristic for sparser star inequalities presented by Caprara *et al.* [7].

We have presented all established y-constraints that will be used to enrich the binary distance model in Chapter 2. We will now continue to describe all linear programming bounds that are known so far. There are two straightforward integer programming formulations. Both use two different types of variables.

Assignment Variables Formulation

The first approach uses **assignment variables** x_{ip} that are equal to 1 if and only if node i is placed in position p, i.e., if $\pi(i) = p$. Additionally, the integral distance variables y according to this ordering are used. The mixed integer programming formulation of the MinLA with these variables is stated as follows.

$$\min \sum_{ij \in E} c_{ij} y_{ij}$$

$$\text{s.t.} \quad \sum_{p=1}^{n} x_{ip} = 1 \quad \text{for all } i \in V, \tag{1.9}$$

$$\sum_{i \in V} x_{ip} = 1 \quad \text{for all } p = 1,\ldots,n, \tag{1.10}$$

$$(x_{ip} + x_{jq} - 1)|p - q| \le y_{ij} \quad \text{for all } ij \in E \text{ for all } p \ne q \in V, \tag{1.11}$$

$$0 \le x_{ij} \le 1 \quad \text{for all } i < j \in V, \tag{1.12}$$

$$y_{ij} \in \mathbb{R}_+ \quad \text{for all } i < j \in V. \tag{1.13}$$

The y-variables could be declared as integer variables, but the formulation remains valid if they are declared as continuous. Constraints (1.9) state that each node is assigned to exactly one position. The restriction that in every position exactly one node can be placed is realized in Constraints (1.10). The correct lower bounds of the integral distance variables y is formulated in Constraints (1.11). If node i is placed in position p and node j in position q, the distance of nodes i and j is exactly $|p - q|$.

It was observed by Caprara *et al.* [7] that the linear programming relaxation admits the trivial solution $x_{ij} = 1/n$ for all $i,j \in V$ and $y_{ij} = 0$ for all edges $ij \in E$. Furthermore, two different types of variables and an enormous number of constraints are used. Therefore this model is of no practical use. Nevertheless, we investigate this model by strengthening several inequalities formulated on integral distance variables only. The results can be found in Section 2.3.

Linear Ordering Approach

The second intuitive way to formulate the MinLA with linear constraints is to use **linear ordering variables**

$$x_{ij} := \begin{cases} 1, & \pi(i) < \pi(j), \\ 0, & \text{otherwise} \end{cases} \quad \text{for all } i < j \in V$$

1.5. Computation of Lower Bounds

to determine the permutation π. Again we need the integral distance variables y to formulate the objective function value. A linear programming formulation with these variables is the following:

$$\min \sum_{ij \in E} c_{ij} y_{ij}$$

$$\begin{aligned}
\text{s.t.} \quad & x_{ij} + x_{ji} = 1 && \text{for all } i < j \in V, & (1.14) \\
& x_{ij} + x_{jl} + x_{li} \leq 2 && \text{for all } i \neq j, l \in V, & (1.15) \\
& \sum_{l=1}^{n}(x_{li} - x_{lj}) \leq y_{ij} && \text{for all } i < j \in V, & (1.16) \\
& \sum_{l=1}^{n}(x_{lj} - x_{li}) \leq y_{ij} && \text{for all } i < j \in V, & (1.17) \\
& 0 \leq x_{ij} \leq 1 && \text{for all } i \neq j \in V, \\
& 1 \leq y_{ij} \leq n-1 && \text{for all } i < j \in V.
\end{aligned}$$

Constraints (1.14) and (1.15) are the tournament and 3-dicycle linear ordering constraints. As the position of a node j within a linear ordering is given by $\deg_{in}(j)+1$ the integral distance variables y can be determined by the x-variables. To be more precise we have

$$y_{ij} = \sum_{l=1}^{n} |x_{li} - x_{lj}| \quad \text{for all } i, j \in V.$$

As this is no linear transformation we cannot formulate the objective function in linear ordering variables only. But for a minimization problem with non-negative edge weights c_{ij}, which is the case for the MinLA, the connection of the two variables can be modeled with constraints (1.16) and (1.17). They require y_{ij} to be the distance of node i and j with respect to the ordering.

It is clear that the lower bound obtained by this optimization is equal to the number of edges of the graph. This is due to the fact that the solution, in which all linear ordering variables are set to $\frac{1}{2}$, is feasible and minimizes the y-variables. Hence, all y_{ij} take their value at the lower bound, which leads to a very bad lower bound for the MinLA. We therefore do not consider this model anymore, but mention it here for the sake of completeness.

Spreading Metric Formulation

The first known suitable integer programming model for the MinLA was formulated by Even et al. [5]. The linear programming relaxation looks like this:

$$\min \sum_{ij \in E} c_{ij} y_{ij}$$

$$\begin{aligned}
\text{s.t.} \quad & \sum_{j \in S} \text{dist}(i,j) \geq \tfrac{1}{4}(|S|^2 - 1) && \text{for all } S \subseteq V \text{ and } i \in S, & (1.18) \\
& y \in \mathbb{R}_+^m,
\end{aligned}$$

where $\text{dist}(i,j)$ is the weight of the shortest path between i and j. The edge weights are given by the node distances y_{st}. Constraint (1.18) is called **spreading constraint**, all feasible solutions of the above formulation are **spreading metrics**. As the violation of the spreading constraint can efficiently be checked using shortest paths methods, this linear programming relaxation can be solved in polynomial time. Another advantage of this formulation is that it is sparse, i.e., it uses only $|E|$ variables. Furthermore, the integrality gap corresponding to this model has become important for the MinLA.

Bornstein & Vempala [89] formulated an alternative linear programming relaxation to compute spreading metrics. But although the formulation is polynomial, the number of variables $O(n^4)$ and number of constraints $O(n^3)$ are high.

Rank Constraint Approach

The **rank of a graph** $\text{rank}(G)$ is defined as the best possible linear arrangement of the unweighted graph G, i. e.,

$$\text{rank}(G) := \min_{\pi \in S(n)} \sum_{ij \in E} |\pi(i) - \pi(j)|.$$

Naturally the determination of the rank is as hard as the solution of MinLA. Nevertheless, Liu & Vannelli [6] formulated a linear program which is based on the computation of the ranks of subgraphs, whose unweighted linear arrangement can be determined in polynomial time.

$$\min \sum_{ij \in E} c_{ij} \, y_{ij}$$

s.t. $\quad \sum_{ij \in E(G')} y_{ij} \geq \text{rank}(G') \qquad G'$ subgraph of G,

$$y_{ij} \geq 1 \qquad \text{for all } i < j \in V,$$

where G' is an ordinary star, a star of node disjoint paths, a k-wheel or a k-prism. Note: In contrast to the spreading metric formulation, this formulation is dense, i. e., it is based on all variables y_{ij} for $i, j \in V$.

Combined Sparse and Dense Integral Distance Variables y Formulation

Both modeling approaches described above are the basis of the recent work of Caprara *et al.* [7]. The advantages of the sparse spreading metric formulation and the dense rank constraint approach are combined in a successful way. The computational results show that for most benchmark instances the best known solutions are not far from the optimum.

Single Row Facility Layout Approach

The MinLA can be considered as a special case of the Single Row Facility Layout problem. In Amaral & Letchford [45] the corresponding more general class of integer polyhedra is investigated. The convex hull is determined and several families of valid inequalities are derived. The results are used within a cutting plane algorithm. We review some of the results in Section 3.2.1.

Betweenness/Consecutive Ones Formulation

We now present the transformation of the MinLA to the Consecutive Ones problem [98]. Imagine n nodes are linearly arranged with positions in $[n]$. The distance of two nodes equals the number of nodes lying between these nodes plus 1. I. e., we have $|\pi(i) - \pi(j)| = \sum_k x^\pi_{ikj} + 1$, where the **betweenness variables** x^π_{ikj} of a linear arrangement π are defined as

$$x^\pi_{ikj} := \begin{cases} 1, & \pi(i) < \pi(k) < \pi(j) \quad \text{or} \quad \pi(i) > \pi(k) > \pi(j), \\ 0, & \text{otherwise.} \end{cases}$$

We therefore have an "explosion" of y-variables into betweenness variables. To obtain an $\mathcal{IP}$ formulation on the betweenness variables we write them into a binary matrix M. The columns of M correspond to the nodes and the rows of M to the edges of the underlying graph. Let $r(i, j)$ be the row belonging to the edge ij, then the entries of M are defined as

$$\overline{M}_{r(i,j),k} := \begin{cases} 1, & k = i \text{ or } k = j, \\ x^\pi_{ikj}, & \text{otherwise.} \end{cases}$$

Now π is a feasible labeling if and only if the matrix M has the consecutive ones property for rows. That is, there is a permutation of the columns of M (which corresponds to π) such that the 1-entries of M occur consecutively in each row of M. There is a characterization of Tucker [99] for the consecutive ones property who gave five types

1.5. COMPUTATION OF LOWER BOUNDS

of forbidden sub matrices. Based on these forbidden matrices one can derive an $\mathcal{IP}$ formulation for the problem. See [98] for a more detailed discussion of this formulation. Although this approach is based on a large number of variables (n^3), the computational results of Caprara et al. [8] are comparable to those of [7].

1.5.3 OTHER BOUNDS

QUADRATIC LINEAR ORDERING

In [100] a different formulation with linear ordering variables is presented. The idea is to optimize non-linear functions over well-studied polytopes, such as the Linear Ordering polytope P_{LO} [101] or the Traveling Salesman polyhedron. The detailed polyhedron studies can then be replaced by the application of some general separation routines for a branch-and-cut algorithm.

Recall that the linear ordering variables x are binary and have the value 1 if and only if node i is placed before node j. The MinLA can, up to a constant, be formulated in the following way, whereas the basic modeling idea is the same as in the consecutive ones formulation.

$$\min \sum_{\substack{i \neq j \neq k \\ i \neq k}} c_{ij}\, x_{ik} x_{kj}$$

$$\text{s.t.} \quad x \in P_{LO},$$

$$x_{ij} \in \{0,1\} \quad \text{for all } i < j \in V.$$

Important for the use of this modeling approach is that only the products of type $x_{ik}x_{kj}$ are necessary, these are $O(n^3)$.

SEMIDEFINITE PROGRAMMING LOWER BOUND

This lower bound is similar to the spreading metric formulation described above. Only two additional constraints are considered, one of them is non-linear.

$$y_{ij} = \|i - j\|_2^2 \quad \text{for all } i < j \in V \qquad (1.19)$$

$$y_{il} \leq y_{ij} + y_{jl} \quad \text{for all } i,j,l \in V. \qquad (1.20)$$

Constraints (1.19) are equivalent to the request that the metric $y \in \mathbb{R}_+^{\binom{n}{2}}$ can be embedded in the so-called negative-type cone, see [95]. Conditions (1.20) are the well-known 3-cycle inequalities. This formulation is again important to prove the integrality gap of the MinLA. A detailed investigation of this model can be found in [91, 92].

EIGENVALUE LOWER BOUND

As the name suggests this bound is based on Eigenvalues. It was introduced by [61] and belongs to the tighter lower bounds of the MinLA, see [73]. Consider the following Laplacian matrix $L_G \in \mathbb{Z}^{n \times n}$ corresponding to the connected graph G.

$$(L_G)_{ij} := \begin{cases} -1, & ij \in E, \\ \deg(i), & i = j, \\ 0, & \text{otherwise.} \end{cases}$$

As L_G is positive semi-definite its smallest Eigenvalue is 0. Therefore we consider the second smallest Eigenvalue λ_2 of L_G. Juvan & Mohar [61] proved that $\lceil \lambda_2(n-1)^2/6 \rceil$ is a lower bound for the MinLA. The Eigenvector to λ_2 is considered in the so-called spectral sequencing heuristic, see Section 1.4.1 on page 22 and Table 1.2 on page 23.

CHAPTER 2

BINARY DISTANCE MODEL

We will now focus on the binary distance model in its complete and sparse version. We start with the inequalities needed for the integer programming formulation and characterize the binary distance model with respect to the similarity and in contrast to the integral y-variables formulation presented in Section 1.5.2 on page 25. Further inequalities are presented and their strength is compared to corresponding y-constraints.

The model is investigated in a second step, in which we consider only those variables for which an edge in the graph exists. We show how the system of constraints must be modified. As we want to obtain a similar strength of the formulation as in the complete case, we present different approaches that help to improve the quality of the formulation.

We then consider an improved mixed linear programming formulation of the MinLA that uses the $\binom{n}{2}$ integral distance variables y together with n^2 binary assignment variables. This idea was developed in cooperation with A. Letchford, Lancaster University, UK.

2.1 BINARY DISTANCE MODEL

2.1.1 DEFINITION AND BASIC PROPERTIES

We present a binary distance modeling approach within a branch-and-cut algorithm for solving linear arrangement problems to proven optimality. The key idea is to introduce binary distance variables d_{ijk} for $1 \leq i < j \leq n$ and $1 \leq k \leq n-1$, where $d_{ijk} = 1$ if nodes i and j have distance k, i.e.,

$$d_{ijk} := \begin{cases} 1, & |\pi(i) - \pi(j)| = k, \\ 0, & \text{otherwise} \end{cases} \quad \text{for all } i < j \in V \text{ and } k = 1, \ldots, n-1.$$

Although this modeling approach has $O(n^3)$ variables, we have several advantages. The d-variables are binary and an integer programming formulation can be given, which is not possible with y-variables. Furthermore we can express every y-constraint for the MinLA problem with d-variables. The well-known distance variables $y_{ij} := |\pi(i) - \pi(j)|$ for all $i < j \in V$ have been widely used for modeling the MinLA, recall Section 1.5.2. These y-variables are in fact an aggregation of the d-variables. To be more precise, the following equation holds:

$$y_{ij} = \sum_{k=1}^{n-1} k \, d_{ijk} \quad \text{for all } i, j \in V. \tag{2.1}$$

The objective function of the MinLA, $\min c^T y$, can be formulated as

$$\min \sum_{ij \in E} c_{ij} \left(\sum_{k=1}^{n-1} k \, d_{ijk} \right).$$

With such a close relationship between these variable types one might wonder whether it is worth working with such similar variables. But a justification of the binary distance model is given in Table 5.2 of Chapter 5. It shows that the gap can be closed between 43% and 53% using the binary d-variables in addition to the integral y-variables. These results are in correspondence with our intuition. While the constraints formulated on y-variables describe the overall structure, the d-variable can realize very specific restrictions. Therefore the combination of y and d-constraints seems promising, as the coarse and the fine structure of the problem can be considered.

Another advantage of the binary distance model is that it can easily be generalized to two dimensions. This is investigated in detail in Wiesberg [47]. Most of the constraints presented in this chapter can be adopted to the two-dimensional case. In particular this holds for the forbidden subgraphs constraints, which is not possible, if the y-variables are refined by betweenness variables, compare Section 1.5.2, instead of our binary distance variables d.

We will now present the most important constraints for the binary distance modeling approach.

$$\sum_{k=1}^{n-1} d_{ijk} = 1 \quad \text{for all } i < j \in V, \tag{2.2}$$

$$\sum_{i<j\in V} d_{ijk} = n - k \quad \text{for all } k = 1, \ldots, n-1, \tag{2.3}$$

$$\sum_{j\neq i} \left(d_{ijk} + d_{ij(n-k)}\right) = 2 \quad \text{for all } i \in V \text{ and } k < \left\lfloor \frac{n}{2} \right\rfloor, \tag{2.4}$$

$$\sum_{j\neq i} d_{ijk} \leq 1 \quad \text{for all } i \in V, k = \left\lfloor \frac{n-1}{2} \right\rfloor + 1, \ldots, n-1, \tag{2.5}$$

$$d_{ijk} \geq 0 \quad \text{for all } i < j \in V \text{ and } k = 1, \ldots, n-1. \tag{2.6}$$

Equations (2.2) state that there is exactly one distance between two nodes. We call them **each-edge-one-distance** equation. Constraints (2.3) specify the connection between the number of pairs with a certain distance and the distance itself, i.e., the distance k occurs exactly $n - k$ times. They are called **the-longer-the-rarer** equation. In Equation (2.4) can be seen that if there are two nodes of small distance k, then there is no distance $n - k$. If, for example, n is odd, j the middle node and $k \leq (n-1)/2$, then there are exactly two edges of length k starting at j: One to the left and one to the right side. If k is bigger, both distances are on one side of the node. We call them **special-degree** equation. Constraints (2.5) are called **degree-big** inequalities. They describe that there is at most one long distance from a node. As the upper bounds for the d-variables are implicitly contained in Constraints (2.2), only the lower bounds (2.6) have to be stated explicitly.

Before we show that the presented d-constraints are an integer programming formulation for the MinLA problem, we state the following.

Remark 2.1.1. *For the sake of simplicity we mention the following constraints.*

$$\sum_{k=1}^{n-1} k(d_{ijk} + d_{jlk} - d_{ilk}) \geq 0 \quad \text{for all } i < j < l \in V, \tag{2.7}$$

$$\sum_{i<j\in V} \sum_{k=1}^{n-1} k d_{ijk} = \binom{n+1}{3}, \tag{2.8}$$

$$1 \leq \sum_{j\neq i} d_{ijk} \leq 2 \quad \text{for all } i \in V, k = 1, \ldots, \lfloor (n-1)/2 \rfloor. \tag{2.9}$$

Constraints (2.7) are the triangle inequalities $y_{ij} + y_{jl} - y_{il} \geq 0$. Equation (2.8) can be obtained as a sum of Equations (2.3). Constraints (2.9) are combinations of Equations (2.4) and (2.5).

2.1.2 Integer Programming Formulation

Proposition 2.1.2. *Constraints (2.2), (2.3), (2.4), (2.5), (2.6) and (2.7) are an integer programming formulation for the MinLA problem.*

2.1. BINARY DISTANCE MODEL

Proof. Assume a solution d for the above stated system of constraints is given. Furthermore assume it is integral. Note: Constraints (2.6) assure that all variable values are non-negative. We will proof that d's incidence vector corresponds to a permutation.

Due to (2.3) there is only one distance of length $n - 1$. Let $l, m \in V$ be the nodes, such that $d_{lm(n-1)} = 1$. Assume w.l.o.g. that l is an inner node, i.e., the longest distance is not between the two outermost nodes of the ordering. We show that this contradicts (2.4). As l is an inner node, there are exactly two edges starting from l having length 1. Otherwise we would not have enough edges of length 1 in Constraints (2.3). Now consider (2.4) with $i = l$, i.e., $\sum_{j \neq l} d_{lj1} + d_{lj(n-1)}$ and note that $d_{lm(n-1)}$ occurs in this sum. Hence, as l is an inner node, $\sum_{j \neq l} d_{lj1} + d_{lj(n-1)} = 3$, which is impossible because of (2.4). Therefore l and m have to be the leftmost and rightmost node within the linear arrangement. We will now construct the permutation from both sides using a certain property of the triangle inequalities, which will be proven now. With constraints (2.8) and (2.7) we obtain

$$\binom{n+1}{3} = \sum_{k=1}^{n-1} k\, d_{lmk} + \sum_{i \notin \{l,m\}} \sum_{k=1}^{n-1} (k\, d_{lik} + k\, d_{imk}) + \sum_{i,j \notin \{l,m\}} \sum_{k=1}^{n-1} k\, d_{ijk}$$

$$= n - 1 + \sum_{i \notin \{l,m\}} \sum_{k=1}^{n-1} (k\, d_{lik} + k\, d_{imk}) + \sum_{i,j \notin \{l,m\}} \sum_{k=1}^{n-1} k\, d_{ijk}$$

$$\geq n - 1 + \sum_{i \notin \{l,m\}} \sum_{k=1}^{n-1} k\, d_{lmk} + \sum_{i,j \notin \{l,m\}} \sum_{k=1}^{n-1} k\, d_{ijk}$$

$$= n - 1 + (n-2)(n-1) + \sum_{i,j \notin \{l,m\}} \sum_{k=1}^{n-1} k\, d_{ijk}$$

$$= n - 1 + (n-2)(n-1) + \binom{n-1}{3}$$

$$= \binom{n+1}{3}.$$

Hence all triangle inequalities involving l and m are tight, i.e.,

$$\sum_{k=1}^{n-1} k(d_{lik} + d_{imk} - d_{mlk}) = 0 \quad \text{for all } i \in V. \tag{2.10}$$

Considering constraints (2.9) we know that for all $k \in \{1, \ldots, \lfloor (n-1)/2 \rfloor\}$ there exists a node $j \in V$ such that $d_{ljk} = 1$. Due to Constraints (2.2) the distance between l and j is unique. For $i = l$ expression (2.10) assures that the distance from j to the rightmost node m is correct. Because of Constraints (2.9) it is possible that there are two different nodes j_1, j_2 having the same distance k to node l. But in this case there would be more than $n - k$ distances of length k within the whole arrangement, which contradicts constraints (2.3).

As the same arguments hold for the rightmost node m, the right half can be determined, too. We obtain an ordering of the n nodes in which each distance k occurs exactly once from each of the outermost nodes l and m.

It remains to be shown that all inner nodes have correct distances. Assume there exists a pair of nodes i', j' having a too large distance. Because of Equation (2.8), there has to exist a pair i, j whose distance is too small. With expression (2.10) we conclude

$$n - 1 > \sum_{k=1}^{n-1} k\, (d_{lik} + d_{ijk} + d_{jmk}) \geq \sum_{k=1}^{n-1} k\, (d_{lik} + d_{imk}) = \sum_{k=1}^{n-1} k\, d_{lmk} = n - 1,$$

which is a contradiction. Therefore all distances within the linear arrangement are correct. □

2.1.3 Further Inequalities

$$\sum_{i,j \in S} d_{ijk} \leq |S| - 1 \qquad \text{for all } S \subset V \text{ for all } k$$

$$\text{such that } |S| - 1 < n - k, \tag{2.11}$$

$$\sum_{j \neq i} \left(d_{ijk} - d_{il(k+1)} \right) \geq 0 \qquad \text{for all } i \in V \text{ and } k = 1, \ldots, \left\lfloor \frac{n-1}{2} \right\rfloor, \tag{2.12}$$

$$\sum_{\substack{j \in V \\ j \neq i}} \sum_{k=1}^{n-1} d_{ijk} \geq N(i) \qquad \text{for all } i \in V. \tag{2.13}$$

Constraints (2.11) are the Equation (2.3) formulated on subsets instead of the whole vertex set. We call them **subtour** inequalities. For $|S| = 3$ they are triangle inequalities, for $|S| > n - k$ they are dominated by (2.3). Constraints (2.12) show that at least as many edges of length k exist from a node i as of length $k+1$. This is due to the fact that if an edge of length k reaches from i to the rightmost node, no edge of length $k+1$ can start from i into the same direction. They are called **monotonic** inequalities. Constraints (2.13) assure that for each node i there have to be as many non-zero variables as there are adjacent nodes. We refer to them as **single-degree** inequalities.

Forbidden Subgraphs

In this section we present some constraints that forbid certain subgraphs. The first inequality of this type is similar to the triangle constraint.

$$d_{ijk_1} + d_{jlk_2} + d_{ilk_3} \leq 2 \qquad \text{for all } i < j < l \in V, \text{ where} \tag{2.14}$$

k_1, k_2 and $k_3 = 1, \ldots, n-1$ are triples of impossible distances on a triangle. We call these constraints (2.14) **special-triangle** inequalities. In Section 2.1.3 we compare the strength of these to the well-known triangle inequalities.

The key idea in the following is to forbid a certain subgraph F in the current LP solution d^*. All inequalities presented now have the same common structure.

$$\sum_{ij \in F} d_{ij\alpha(i,j)} \leq |F| - 1 \qquad \text{for every forbidden subgraph } F = (V, E, \alpha),$$

where α denotes the distances between nodes of F with respect to d^*. In a first step we consider forbidden subgraphs F that consist of short edges connecting two 1-paths. A 3-**bridge**, see Figure 2.1-(a), consists of five nodes and four edges of length 1. It is impossible to embed this subgraph within a feasible linear arrangement. In a 4-**bridge** we have six nodes, four edges of length 1 and one edge of length ≤ 2 that connects the 1-paths, see Figure 2.1-(b). When the connecting edge has length ≤ 3 and one of the 1-paths contains five nodes, we call this subgraph a 6-**bridge**. In the case of a connecting path of length ≤ 4 and two 1-paths of length 4, we talk of an 8-**bridge**. The last two subgraphs are shown in Figure 2.1-(c) and (d). The corresponding **bridge** inequalities are formulated in (2.15) to (2.18).

$$\sum_{ij \in B_3} d_{ij\alpha(i,j)} \leq 3 \qquad \text{for every 3-bridge } B_3 = (V, E, \alpha), \tag{2.15}$$

$$\sum_{ij \in B_4} d_{ij\alpha(i,j)} \leq 4 \qquad \text{for every 4-bridge } B_4 = (V, E, \alpha), \tag{2.16}$$

$$\sum_{ij \in B_6} d_{ij\alpha(i,j)} \leq 6 \qquad \text{for every 6-bridge } B_6 = (V, E, \alpha), \tag{2.17}$$

$$\sum_{ij \in B_8} d_{ij\alpha(i,j)} \leq 8 \qquad \text{for every 8-bridge } B_8 = (V, E, \alpha), \tag{2.18}$$

where $\alpha(i,j)$ is the distance of edge ij in the bridge with respect to d^*.

The same constraint idea can be applied to large star subgraphs displayed in Figure 2.2, which we call **path-stars**. The **path-star** inequalities are formulated as follows.

2.1. BINARY DISTANCE MODEL

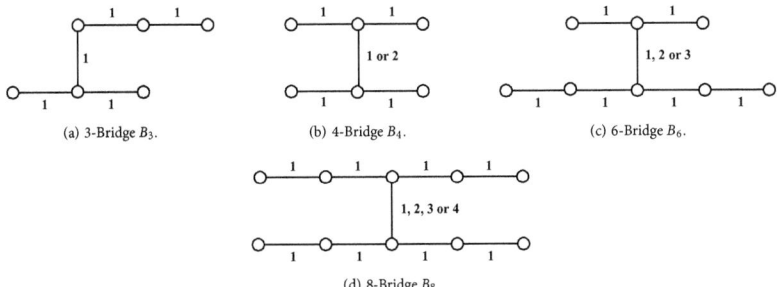

Figure 2.1: Several forbidden subgraphs of bridge type.

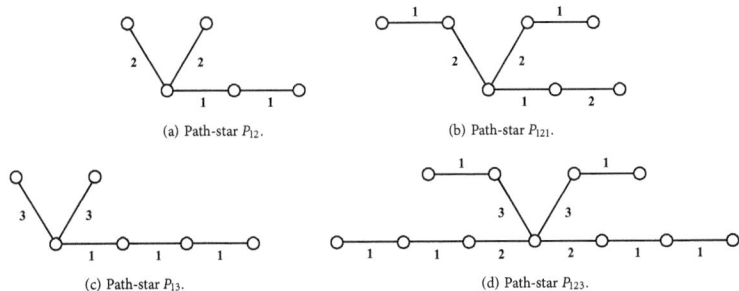

Figure 2.2: Several forbidden subgraphs of path-star type.

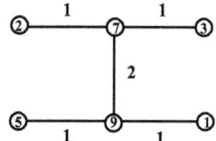

Figure 2.3: Example of a 4-bridge within a current LP solution.

$$\sum_{ij \in P_{12}} d_{ij\alpha(i,j)} \leq 3 \quad \text{for every path-star } P_{12} = (V,E,\alpha), \tag{2.19}$$

$$\sum_{ij \in P_{121}} d_{ij\alpha(i,j)} \leq 5 \quad \text{for every path-star } P_{121} = (V,E,\alpha), \tag{2.20}$$

$$\sum_{ij \in P_{13}} d_{ij\alpha(i,j)} \leq 4 \quad \text{for every path-star } P_{13} = (V,E,\alpha), \tag{2.21}$$

$$\sum_{ij \in P_{123}} d_{ij\alpha(i,j)} \leq 9 \quad \text{for every path-star } P_{123} = (V,E,\alpha). \tag{2.22}$$

One might wonder why we do not formulate these constraints as rank inequalities. We therefore present a comparison of rank constraints and forbidden subgraph constraints.

FORBIDDEN SUBGRAPHS VERSUS RANK CONSTRAINTS

Consider the rank constraint

$$\sum_{ij \in E} y_{ij} \geq 7 \quad \text{for all 4-bridges } B_4 = (V,E) \tag{2.23}$$

corresponding to (2.16). It is clear that rank constraints are more general than constraints forbidding certain subgraphs within an LP solution. Therefore we formulate as many constraints as possible as rank constraints instead of forbidden subgraph inequalities. Nevertheless it turned out that the the rank constraints corresponding to the forbidden subgraphs are too weak. This is due to the fact that rank constraints are formulated on the integral distance variables y. With these variables it is not possible to forbid subgraphs with certain, infeasible distances. An example for such a case is the LP solution $d^*_{2\,7\,1} = d^*_{7\,3\,1} = d^*_{5\,9\,1} = d^*_{1\,9\,1} = 1$ and $d^*_{7\,9\,2} = 0.8$, see Figure 2.3. Hence constraint (2.16) is violated but (2.23) is not, as other variables $d^*_{7\,9\,8} = 0.175$ and $d^*_{7\,9\,9} = 0.025$ compensate the mistake.

COMPARISON OF DIFFERENT TRIANGLE INEQUALITIES

Due to their similarity we investigate the correlation between the y- and d triangle inequalities

$$y_{ij} + y_{jl} - y_{il} \geq 0 \quad \text{for all } i < j < l \in V, \tag{2.24}$$

$$d_{ijk_1} + d_{jlk_2} + d_{ilk_3} \leq 2 \quad \text{for all } i < j < l \in V, \tag{2.25}$$

where k_1, k_2 and $k_3 = 1,\ldots,n-1$ are triples of impossible distances on a triangle. It would be important to know whether one type of inequality is dominated by the other one. Our tests show that this is not the case. In fact we will give two example LP solutions d^* in which one constraint type is violated and the other one is not. Consider

$$d^*_{2\,9\,2} = 0.25, d^*_{2\,9\,3} = 0.5 \quad \Rightarrow y^*_{2\,9} = 2,$$
$$d^*_{9\,4\,1} = 0.25, d^*_{9\,4\,3} = 0.25 \quad \Rightarrow y^*_{9\,4} = 1, \quad \text{and}$$
$$d^*_{2\,4\,8} = 0.25, d^*_{2\,4\,4} = 0.5 \quad \Rightarrow y^*_{2\,4} = 4.$$

The triangle inequality $y^*_{2\,9} + y^*_{9\,4} - y^*_{2\,4} = -1 \not\geq 0$ is violated. In contrast no d triangle is violated as all d^* values are so small. No sum of d variables with the distances $k_1, k_2, k_3 = 1,\ldots, n-1$ of impossible distances on a triangle is larger than 2.

2.1. Binary Distance Model

On the other hand consider the following LP solution.

$$d^*_{3\,5\,3} = 0.6, d^*_{3\,5\,2} = 0.1 \Rightarrow y^*_{3\,5} = 2,$$
$$d^*_{5\,8\,3} = 0.6, d^*_{5\,8\,2} = 0.1 \Rightarrow y^*_{5\,8} = 2, \text{ and}$$
$$d^*_{3\,8\,4} = 0.9, d^*_{3\,8\,2} = 0.2 \Rightarrow y^*_{3\,8} = 4.$$

The triangle $y^*_{3\,5} + y^*_{5\,8} - y^*_{3\,8} = 0$ is not violated, but for $k_1 = 3$, $k_2 = 3$ and $k_3 = 4$ the special triangle constraint $d^*_{3\,5\,k_1} + d^*_{5\,8\,k_2} - y^*_{3\,8\,k_3} = 2.1 \not\leq 2$ is violated.

2.1.4 Rank Constraints on y-Variables

In our branch-and-cut-and-price algorithm we focus on d-constraints but use the following y-inequalities, recall Section 1.5.2 for details.

- (Sparser) Star inequalities
- Clique inequalities
- Cycle inequalities
- Wheel inequalities
- Prism inequalities
- (Large) Hypermetric inequalities including the special case triangle inequalities
- (Large) Bipartite inequalities

Furthermore we introduce the following new rank constraints.

$$\sum_{ij \in E} y_{ij} \geq c_{\text{opt}}(G') \quad \text{for every subgraph } G' = (V,E) \text{ of } G, \tag{2.26}$$

$$\sum_{ij \in E} y_{ij} \geq c_{\text{opt}}(D) \quad \text{for every degree-subset } D = (V,E) \text{ of } G, \tag{2.27}$$

$$\sum_{ij \in E} y_{ij} \geq 2|V| + 2 \quad \text{for every diamond } (V,E), \tag{2.28}$$

$$\sum_{ij \in E} y_{ij} \geq 9 \quad \text{for every 3-cycle-star } (V,E), \tag{2.29}$$

$$\sum_{ij \in E} y_{ij} \geq 14 \quad \text{for every 4-cycle-star } (V,E), \tag{2.30}$$

$$\sum_{ij \in E} y_{ij} \geq 7 \quad \text{for every 3-cycle-with-2-legs } (V,E), \tag{2.31}$$

$$\sum_{ij \in E} y_{ij} \geq 10 \quad \text{for every 3-cycle-with-4-legs } (V,E), \tag{2.32}$$

$$\sum_{ij \in E} y_{ij} \geq 18 \quad \text{for every 3-cycle-with-6-legs } (V,E), \tag{2.33}$$

$$\sum_{ij \in E} y_{ij} \geq 9 \quad \text{for every 4-cycle-with-2-legs } (V,E), \tag{2.34}$$

$$\sum_{ij \in E} y_{ij} \geq 12 \quad \text{for every 4-cycle-with-4-legs } (V,E) \text{ of type } A, \tag{2.35}$$

$$\sum_{ij \in E} y_{ij} \geq 16 \quad \text{for every 4-cycle-with-4-legs } (V,E) \text{ of type } B. \tag{2.36}$$

The subgraphs G' in the so-called **subgraph** inequalities (2.26) are stars that contain all existing edges between the neighbors of the center node. As the MinLA problem of G' should be solvable in reasonable time we consider at

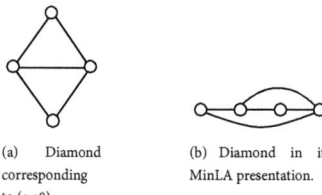

(a) Diamond corresponding to (2.28).

(b) Diamond in its MinLA presentation.

Figure 2.4: Diamond and its MinLA representation.

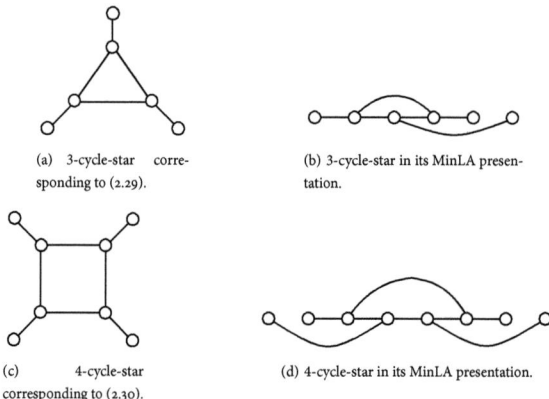

(a) 3-cycle-star corresponding to (2.29).

(b) 3-cycle-star in its MinLA presentation.

(c) 4-cycle-star corresponding to (2.30).

(d) 4-cycle-star in its MinLA presentation.

Figure 2.5: Cycle-stars and their MinLA representation.

most 8 neighbors of the center node. As we observed that nodes with high degrees are often close together within an optimal solution we want to separate on subsets consisting of nodes having large degrees. We therefore order the nodes of the graph G by their degree and then consider the subgraphs D defined by all edges of the first eight nodes, the nodes 2 to 9, 3 to 10 and so forth. We call these the **degree-subset** inequalities, they are presented in (2.27). The **diamond** inequalities (2.28) are rank constraints on diamonds, were a diamond is shown in Figure 2.4. For the 3- and 4-cycle-stars, see Figure 2.5, we formulated the rank constraints (2.29) and (2.30). We call them 3-/4-**cycle-star** inequalities. Constraints (2.31) to (2.36) are rank constraints for 3- and 4-cycle with 2 or 4 legs. These cycle-like structures are displayed in Figures 2.6 and 2.7 and are called 3-/4-**cycle-with-legs**. The corresponding inequalities are denoted by 3-/4-**cycle-with-legs** inequalities.

2.1. BINARY DISTANCE MODEL

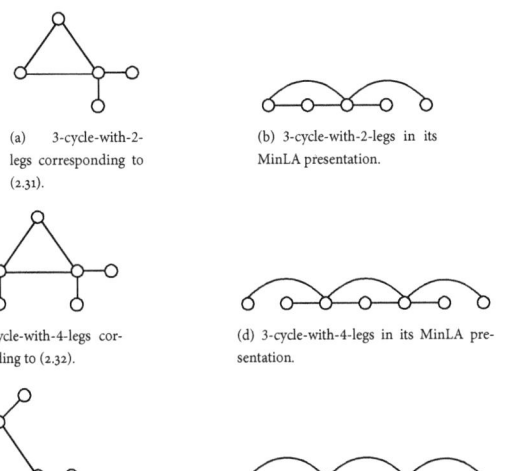

(a) 3-cycle-with-2-legs corresponding to (2.31).

(b) 3-cycle-with-2-legs in its MinLA presentation.

(c) 3-cycle-with-4-legs corresponding to (2.32).

(d) 3-cycle-with-4-legs in its MinLA presentation.

(e) 3-cycle-with-6-legs corresponding to (2.33).

(f) 3-cycle-with-6-legs in its MinLA presentation.

Figure 2.6: 3-cycle-with-legs and their MinLA representation.

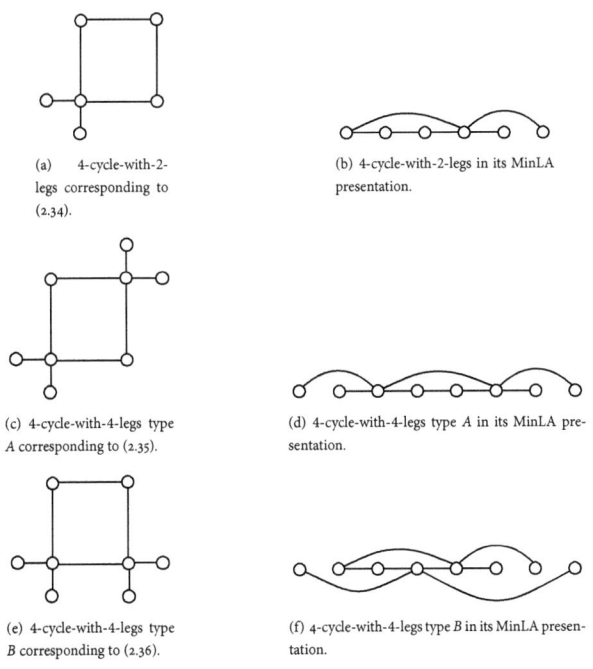

(a) 4-cycle-with-2-legs corresponding to (2.34).

(b) 4-cycle-with-2-legs in its MinLA presentation.

(c) 4-cycle-with-4-legs type A corresponding to (2.35).

(d) 4-cycle-with-4-legs type A in its MinLA presentation.

(e) 4-cycle-with-4-legs type B corresponding to (2.36).

(f) 4-cycle-with-4-legs type B in its MinLA presentation.

Figure 2.7: 4-cycle-with-legs and their MinLA representation.

2.2 Sparse Problem Formulation with Binary Distance Variables

We will now consider the sparse problem formulation of the binary distance model. I. e., we do not consider the d-variables for all $i, j \in V$ but only for edges $ij \in E$ of the underlying graph. That reduces the number of variables from $O(n^3)$ to $|E|(n-1)$. As most of the benchmark graphs have a sparse structure this is a strong decrease in the number of variables.

2.2.1 Basic Properties and Modified Constraint System

As a consequence the system of constraints gets significantly smaller which is important for the practical use of the model. The system has several modifications which are highlighted by bold writing.

$$\sum_{k=1}^{n-1} d_{ijk} = 1 \quad \text{for all } ij \in E, \tag{2.37}$$

$$\sum_{ij \in E} d_{ijk} \leq n - k \quad \text{for all } k = 1, \ldots, n-1, \tag{2.38}$$

$$\sum_{j \neq i, ij \in E} \left(d_{ijk} + d_{ij(n-k)}\right) \leq 2 \quad \text{for all } i < n-1 \text{ and } k < \left\lfloor \frac{n}{2} \right\rfloor, \tag{2.39}$$

$$\sum_{j \neq i, ij \in E} d_{ijk} \leq 1 \quad \text{for all } i \in V, k = \left\lfloor \frac{n-1}{2} \right\rfloor + 1, \ldots, n-1, \tag{2.40}$$

$$\sum_{ij \in S(E)} d_{ijk} \leq |S| - 1 \quad \text{for all } S \subset V \text{ for all } k$$

$$\text{such that } |S| - 1 < n - k, \tag{2.41}$$

$$\sum_{j \in V, j \neq i \atop ij \in E} \sum_{k=1}^{n-1} d_{ijk} \geq N(i) \quad \text{for all } i \in V, \tag{2.42}$$

$$d_{ijk_1} + d_{jlk_2} + d_{ilk_3} \leq 2 \quad \text{for all } ij, jl, il \in E,$$

$$\text{for all non-3-cycles } k_1, k_2, k_3 \leq n - 1, \tag{2.43}$$

$$\sum_{k=1}^{n-1} k(d_{ijk} + d_{jlk} - d_{ilk}) \geq 0 \quad \text{for all } ij, jl, il \in E, \tag{2.44}$$

$$d_{ijk} \geq 0 \quad \text{for all } ij \in E \text{ and } k = 1, \ldots, n-1. \tag{2.45}$$

The first thing to note is that we have less equations of type (2.37). Furthermore, constraint (2.38) is no equation anymore but a lower bound inequality. This is due to the fact that we cannot guarantee the existence of all edges of a certain length. The situation is similar for constraint (2.39), in which we now sum only over existing edges and cannot assure equality anymore. The summation in (2.40) and (2.42) has to be restricted as well. In constraint (2.41) only the edges between nodes of the subset S are considered. They are denoted by $S(E)$. The monotonic inequality (2.12) cannot be adopted to the sparse problem formulation. The two types of triangles (2.43) and (2.44) can only be formulated for existing edges. The same holds for the lower bounds (2.45). Note: The forbidden structure inequalities and rank constraints do not change. Nevertheless, the consequence of the described modifications are not dispensable. On the one hand we cannot give an integer programming formulation anymore. On the other hand the lower bounds get worse, as it can be seen in Section 5.10. We therefore present three different ways to strengthen the sparse problem formulation.

2.2.2 Additional Variables

We enrich the system by adding two different types of variables: Transitive variables and variables to formulate (2.38) as equation. In Figure 2.8 we give an overview of the first type of additional variables. For edges of the underlying graph, displayed in a solid black line, we introduce the artificial transitive edge displayed in a thick grey line, see

2.2. Sparse Problem Formulation

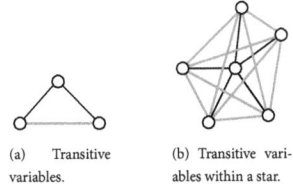

(a) Transitive variables. (b) Transitive variables within a star.

Figure 2.8: Additional variables of transitive type.

Figure 2.8-(a). Note: These transitive variable completions include the addition of all edges within a star. Figure 2.8-(b) presents this particular type of transitive edges. The second type of additional variables are motivated by the wish to "rescue" as much equations from the complete problem formulation as possible. We therefore introduce all variables necessary to satisfy the equation

$$\sum_{i<j\in V} d_{ijk} = n - k \quad \text{for all } k = \lfloor n/2 \rfloor - 2, \ldots, \lfloor n/2 \rfloor + 2.$$

2.2.3 Shortest Path Strengthening

This approach was investigated by Caprara et al. [7] at the same time. Consider the values y^* corresponding to the current LP solution d^*. We compute all shortest paths for $ij \notin E$ with respect to y^*. We then modify the constraints by adding the shortest (i,j)-path for every non-existent y-variable. Let P_{ij} denote this shortest (i,j)-path with respect to y^* and let its length be $y(P_{ij})$. We define

$$d(i,j) := \begin{cases} y_{ij} & \text{if } (i,j) \in E, \\ y(P_{ij}) & \text{otherwise.} \end{cases}$$

If P_{ij} consists of the nodes $w_1 = i, w_2, \ldots, w_k = j$, it implies the upper bound

$$\sum_{l=0}^{k-1} d(w_l, w_{l+1}) \geq y_{ij} \quad \text{for all } i < j \in V.$$

2.2.4 Extended Star Constraint

How the constraints can be extended to V is shown with the example of the star constraint. Let (S, E) be a star with center node r

$$\sum_{i\in S\setminus\{r\}} y_{ir} \geq \lfloor (|S|+1)^2/4 \rfloor. \tag{2.46}$$

Using the shortest path distances the corresponding inequality can be extended to V in the following way. We first extend to V using the shortest path lengths

$$\sum_{i\in V} d(i,r) \geq \lfloor |V|^2/4 \rfloor. \tag{2.47}$$

Then we express the shortest paths in existing variables,

$$\sum_{ij\in E} a_{ij} y_{ij} \geq \lfloor |V|^2/4 \rfloor, \tag{2.48}$$

and obtain the extended star constraint. This procedure can be applied to several constraints. We only have to be careful to keep the modified constraints feasible. Therefore we can only modify $\geq$ inequalities with positive coefficients. In Section 2.2.4 we present the lower bound improvement achieved by the extended star constraint. As the extended constraints tend to be quite dense we do not want to work with a lot of these. Nevertheless, Caprara et al. [7] present a very successful combination of a dense and sparse problem formulation using y-variables.

2.3 Assignment Variables Formulation Revisited

We examine the possibility of formulating the MinLA as an improved version of the Assignment Variables Formulation. Recall this model from Section 1.5.2 on page 26.

$$\min \sum_{ij \in E} c_{ij} y_{ij},$$

s.t.
$$\sum_{p=1}^{n} x_{ip} = 1 \quad \text{for all } i \in V, \tag{2.49}$$

$$\sum_{i \in V} x_{ip} = 1 \quad \text{for all } p = 1, \ldots, n, \tag{2.50}$$

$$y_{ij} \geq (x_{ip} + x_{jq} - 1)|p - q| \quad \text{for all } ij \in E, \text{ for all } p \neq q \in V, \tag{2.51}$$

$$0 \leq x_{ij} \leq 1 \quad \text{for all } i < j \in V, \tag{2.52}$$

$$y_{ij} \in \mathbb{R}_+ \quad \text{for all } i < j \in V. \tag{2.53}$$

Unfortunately, this formulation has $O(n^4)$ constraints, which is excessive.

2.3.1 Improvement of the Formulation

A much more compact formulation can be obtained by replacing constraints (2.51) with the following constraints.

$$y_{ij} \geq \sum_{p=1}^{n} p(x_{ip} - x_{jp}) \quad \text{for all } i < j \in V \text{ and} \tag{2.54}$$

$$y_{ij} \geq \sum_{p=1}^{n} p(x_{jp} - x_{ip}) \quad \text{for all } i < j \in V. \tag{2.55}$$

This improved formulation has only $O(n^2)$ constraints. Its LP relaxation is likely to be very weak, but we can strengthen it. One could of course simply use the valid inequalities that were presented by Amaral & Letchford [45]. More interestingly, however, it is possible to exploit the presence of the additional assignment variables to derive new and/or stronger valid inequalities. This will be done in Chapter 3.

Chapter 3

Polyhedral Theory

In this chapter we address the polyhedral properties of all models considered in this thesis. We start with some preparative definitions in Section 3.1. A short overview of the work on $P_n(G)$, which is the polyhedron corresponding to the y-variables formulation, and on DOM_n, its Minkowski sum with the non-negative orthant of $\mathbb{R}^{|E|}$, is given in Section 3.2. In Section 3.3 we turn our attention to the convex set Q_n. We are particularly interested in its relationship to P_n and to the cut cone CUT_n, in its unbounded edges, and in the unbounded edges of its closure $\overline{Q_n}$. In order to achieve a complete set of results for $\overline{Q_n}$, we display some results on facet-defining inequalities from Letchford [102]. The polyhedron D_n corresponding to the complete problem formulation with d-variables is investigated in Section 3.4. This is followed by a study of the polyhedron P_n^A of the revisited assignment variables formulation. All results of this last section were obtained in cooperation with A. Letchford, Lancaster University, UK, whereas results presented in Section 3.3 were obtained together with A. Letchford and D. O. Theis, Universität Magdeburg, Germany.

3.1 Preparative Definitions

The following facts about fans, zonotopes, and the permutahedron can be found in [17] and [103]. Let L^n be a linear vector space of dimension n. A **fan** in L^n is a family $\mathcal{F} = \{C_1, \ldots, C_k\}$ of nonempty polyhedral cones $C_1, \ldots, C_k$ with the following properties:

- every nonempty face of a cone of $\mathcal{F}$ is a cone of $\mathcal{F}$ itself,
- the intersection of two cones of $\mathcal{F}$ is an element of $\mathcal{F}$ again.

A fan is **complete** if the union of its cones is the whole space. Furthermore, by definition the faces of a fan are its cones. The **face fan** of a polyhedron P is defined by

$$\mathcal{F}(P) := \{\text{cone}(F) \mid F \text{ face of } P, F \neq P\}.$$

For a polyhedron P in a vector space L^n, the set

$$\mathcal{N}(P) := \{N_F \mid F \text{ face of } P, F \neq \emptyset\}, \text{ where}$$

$$N_F := \left\{ c \in (L^n)^* \mid F \subseteq \{x \in P \mid cx = \max_{y \in P} cy\} \right\},$$

is the so-called **normal fan** of the polyhedron P. It is a collection of cones N_F, where F ranges over the nonempty faces of P. For any such F, the cone N_F is defined as the set of all vectors $c \in L^n$ for which the maximum of the linear function $x \mapsto c^T x$ over P is attained in all points of F. The normal fan is a complete fan in the dual space $(L^n)^*$. We will be most interested in the normal fan $N_\pi := N_{\{v_\pi\}}$, where v_π is the vertex of the permutahedron corresponding to the permutation $\pi \in S(n)$, see Equation 3.2.

Theorem 3.1.1 ([17]). *Let P be a polytope with $0 \in \text{int}(P)$. The normal fan of P equals the face fan of its polar, i.e., $\mathcal{N}(P) = \mathcal{F}(P^\triangle)$ and $N_F = \text{cone}(F^\circ)$.*

Proof. Let F be an arbitrary nonempty face of P. Then

$$N_F = \left\{ c \in (L^n)^* \,\middle|\, F \subseteq \left\{ x \in P \,\middle|\, cx = \max_{y \in P} cy \right\} \right\} \quad (3.1)$$

$$= \left\{ \lambda c \,\middle|\, c \in (L^n)^*, \lambda \in \mathbb{R}_+, cx = 1 \text{ for all } x \in F \right\}$$

$$= \text{cone}(F^\circ).$$

$\square$

The finite set $\mathcal{A} := \{H_1, \ldots, H_p\}$ of linear hyperplanes $H_i := \{x \in L^n \mid xv_i = 0\}$ is called a **linear hyperplane arrangement**. It decomposes L^n into a complete fan $\mathcal{F}_\mathcal{A}$, which is the **fan of the linear hyperplane arrangement** $\mathcal{A}$ of L^n.

A **zonotope** is the affine projection of a cube $C^p := [-1,+1]^p = \sum_{k=1}^p [-e_k, e_k]$. That means there exists a linear form $V : L^n \to L^n$, $b \in L^n$ and a mapping $f : x \mapsto Vx + b$ with V being a linear matrix such that $f(C^p) = \sum_{k=1}^p [-v_k, v_k] + b$, where $[x,y]$ is the line segment joining the two points x and y.

Theorem 3.1.2 ([17]). *Let $P \subseteq L^n$ be a zonotope and let $v_1, \ldots, v_p$ be the vectors that determine the line segments whose Minkowski sum is equal to P. It is true that the normal fan $\mathcal{N}(P)$ of P is the fan $\mathcal{F}_\mathcal{A}$ of the linear hyperplane arrangement $\mathcal{A} := \{H_1, \ldots, H_p\}$ of L^n with $H_i := \{c \in (L^n)^* \mid cv_i = 0\}$.*

The **permutahedron** Π_{n-1} is defined as

$$\Pi^{n-1} := \left[-\tfrac{1}{2}(e_2 - e_1), \tfrac{1}{2}(e_2 - e_1) \right] + \left[-\tfrac{1}{2}(e_3 - e_1), \tfrac{1}{2}(e_3 - e_1) \right]$$
$$+ \ldots + \left[-\tfrac{1}{2}(e_n - e_{n-1}), \tfrac{1}{2}(e_n - e_{n-1}) \right].$$

More commonly, the permutahedron is known in its translated form,

$$\Pi^{n-1} + \frac{n+1}{2}\mathbf{1} = \text{conv}\{\pi \mid \pi \text{ is a permutation of the points } [n]\}.$$

We will denote the vertex of Π^{n-1} corresponding to the permutation $\pi \in S(n)$ by

$$v_\pi := \pi - \frac{n+1}{2}\mathbf{1}. \quad (3.2)$$

The permutahedron is a simple zonotope of dimension $n-1$ and the affine projection of the cube of dimension $\binom{n}{2}$. When written in the form we defined it, Π^{n-1} is full dimensional in the linear subspace

$$L^n := \left\{ x \in \mathbb{R}^n \,\middle|\, \sum_{i=1}^n x_i = 0 \right\}$$

of $\mathbb{R}^n$. Moreover, the permutahedron is symmetric: $\Pi^{n-1} = -\Pi^{n-1}$ and $0^T \in \text{int}(\Pi^{n-1})$. This makes Π^{n-1} easier to work with and we can define the polar $(\Pi^{n-1})^\triangle$ of the permutahedron, which is simplicial. Balas characterized in [103] the facet-defining inequalities of $\Pi^{n-1} + \frac{1}{n+1}\mathbf{1}$ to be

$$\sum_{i \in U} x_i \geq \binom{|U|+1}{2}, \quad (3.3)$$

where $\varnothing \neq U \subsetneq [n]$. Hence, facets of the permutahedron correspond to nonempty sets $U \subsetneq [n]$. The permutations $\pi \in S(n)$ with $\sum_{i \in U} \pi(i) = \binom{|U|+1}{2}$ are exactly those with $U = \{\pi^{-1}(1), \ldots, \pi^{-1}(k)\}$ and $|U| = k$. Note that for the antipodal permutation π^- of π, we analogously obtain $U^c = \{\pi^{-1}(1), \ldots, \pi^{-1}(k')\}$ and $|U^c| = k'$. We call a permutation $\pi \in \text{vert}(\Pi^{n-1})$ and a nonempty set $U \subsetneq [n]$ **incident**, if and only if $U = \{\pi^{-1}(1), \ldots, \pi^{-1}(k)\}$ with $|U| = k$. Thus, incidence of permutations and nonempty subsets of $[n]$ reflects incidence of vertices and facets of the permutahedron and, of course, of facets and vertices of the polar of the permutahedron.

A consequence of Theorem 3.1.2 for the permutahedron is the following corollary.

3.1. Preparative Definitions

Corollary 3.1.3 ([17]). *The normal fan of the permutahedron $\mathcal{N}(\Pi^{n-1})$ is equal to the fan of the hyperplane arrangement $\mathcal{F}_\mathcal{A}$, where $\mathcal{A} := \{H_{k,l} \mid 1 \leq k, l \leq n\}$ and*

$$H_{k,l} := \left\{ c \in L^n \mid c(e_k - e_l) = 0 \right\}.$$

□

This result leads to an appropriate description of the cones of the normal fan we are most interested in, namely $N_\pi := N_{\{v_\pi\}}$, see expression (3.1). Define $H^0_{k,l} := H_{k,l}$, $H^+_{k,l} := \{c \in (L^n)^* \mid c(e_k - e_l) > 0\}$ and $H^-_{k,l} := \{c \in (L^n)^* \mid c(e_k - e_l) < 0\}$. Further, set $\mathrm{sign}(\lambda) := +$ for $\lambda \in \mathbb{R}_+$, $\mathrm{sign}(\lambda) := -$ for $\lambda \in \mathbb{R}_-$ and $\mathrm{sign}(0) := 0$.

Corollary 3.1.4. *We have*

$$N_\pi = \left\{ c \in (L^n)^* \mid \mathrm{sign}(c_k - c_l) = \mathrm{sign}(\pi(k) - \pi(l)) \text{ for all } 1 \leq k, l \leq n \right\}.$$

Proof. For $1 \leq k, l \leq n$ it is true that $c_k - c_l > 0 \Leftrightarrow c(e_k - e_l) > 0 \Leftrightarrow c \in H^+_{k,l}$. That means given $\mathrm{sign}(c_k - c_l)$, we know in which side of the hyperplane $H_{k,l}$ the point c is contained. In this way, the position of c with respect to all hyperplanes $H_{k,l}, 1 \leq k, l \leq n$ can be determined. Therefore

$$c \in \bigcap_{k,l=1}^n H^{\mathrm{sign}(c(e_k - e_l))}_{k,l}.$$

An intersection of open half spaces and hyperplanes is nonempty if and only if the ordering of the components of c is consistent. Hence, if the intersection is nonempty, there exists a permutation π which corresponds to the ordering of the components of c. Let C be a facet of $\mathcal{F}_\mathcal{A}$. Then, by Corollary 3.1.3, C can be described as

$$C = \left\{ c \in (L^n)^* \mid c \in \bigcap_{k,l=1}^n H^{\mathrm{sign}(c(e_k - e_l))}_{k,l}, \, \mathrm{sign}(c(e_k - e_l)) \in \{+, -\} \text{ for all } 1 \leq k, l \leq n \right\}$$

$$= \left\{ c \in (L^n)^* \mid \mathrm{sign}(c_k - c_l) = \mathrm{sign}(\pi(k) - \pi(l)) \text{ for all } 1 \leq k, l \leq n \right\}$$

$$= \left\{ c \in (L^n)^* \mid c\pi = \max_{y \in \Pi^{n-1}} cy \right\}$$

$$= N_\pi.$$

□

With this useful description of N_π we conclude.

Corollary 3.1.5. *For all $\pi \in S(n)$ we have $v_\pi \in N(\pi)$.*

Proof. Obviously, from the definitions of v_π and L^n it immediately follows that $v_\pi \in L^n$. As we have $\pi(k) < \pi(l) \Leftrightarrow \pi(k) - \frac{n+1}{2} < \pi(l) - \frac{n+1}{2}$ for all k, l Corollary 3.1.4 and the definition of v_π imply $v_\pi \in N_\pi$. □

Lemma 3.1.6. *For every permutation π we know that N_π is a $(n-1)$-dimensional simplicial cone which has $n-1$ extreme rays and whose apex is 0.*

Proof. From the definition of the cones N_π it is clear that 0 has to be an element of all cones of the normal fan, it is therefore the apex of all. As the permutahedron is simple, each cone in the normal fan is simplicial.

Let π be an arbitrary element of $S(n)$. As v_π is a vertex of Π^{n-1}, from Theorem 3.1.1 it follows that $\{v_\pi\}^\circ$ is a facet of the polar $(\Pi^{n-1})^\triangle$. In particular, $N_\pi = \mathrm{cone}(\{v_\pi\}^\circ)$ and $\dim(N_\pi) = n - 1$ because $\{v_\pi\}^\circ$ is a facet.

As the permutahedron Π^{n-1} is simple, every vertex of Π^{n-1} is contained in exact $n-1$ facets of Π^{n-1}, and by Theorem 3.1.1 we get that $\{v_\pi\}^\circ$ contains exactly $n-1$ vertices of $(\Pi^{n-1})^\triangle$. These vertices of $\{v_\pi\}^\triangle$ correspond exactly to the extreme rays in N_π, hence N_π has $n-1$ extreme rays. □

Proposition 3.1.7. *We have*

$$v_\pi + N_\pi = \{x \in N_\pi \mid |x_k - x_l| \geq 1 \text{ for all } 1 \leq k \neq l \leq n\}.$$

Proof. Let π be any element of $S(n)$. In the first part we show "⊆". Consider $x \in N_\pi$. As $v_\pi \in N_\pi$, by Lemma 3.1.5, we have $x + v_\pi \in N_\pi$ because N_π is a convex cone. For any k, l with $\pi(k) < \pi(l)$, we obtain, as $x_k \le x_l$, that $x_l + v_{\pi l} - (x_k + v_{\pi k}) \ge v_{\pi l} - v_{\pi k} = \pi(l) - \pi(k) \ge 1$. This implies $|x_i + v_{\pi i} - (x_j + v_{\pi j})| \ge 1$ for all $i \ne j$, and hence $x + v_\pi \in R_n$. This proves $v_\pi + x \in R_n \cap N_\pi$.

Now "⊇" has to be shown. Let x be any element of N_π with $|x_k - x_l| \ge 1$. We define $y := x - \pi$ and prove $y \in N_\pi$. As $x \in N_\pi$ holds, $\pi(l) > \pi(k)$ implies $x_l \ge x_k$, and from our assumption we even have $x_l - x_k \ge 1$. Now for $\pi(k) - \pi(l) =: r \ge 1$ there exist $j_0, \ldots, j_r$ with $\pi(k) = \pi(j_0) < \ldots < \pi(j_r) = \pi(l)$, and we can telescope

$$x_k - x_l = \sum_{i=0}^{r-1}(x_{k_{i+1}} - x_{k_i}) \ge \sum_{i=0}^{r-1} 1 = r = \pi(k) - \pi(l).$$

Therefore

$$y_k - y_l = x_k - v_{\pi,k} - (x_l - v_{\pi,l}) = x_k - \pi(k) - (x_l - \pi(l)) \ge 0,$$

and we have shown that for $y, \pi \in N_\pi$, $y - \pi$ is an element of N_π as well. □

Considering Theorem 3.1.1, we obtain that the cones N_π are generated by the points

$$a_U := \frac{2}{n(n-k)} \chi^{U^c} - \frac{2}{kn} \chi^U,$$

where U is a nonempty proper subset of $[n]$ and $k = |U|$. This is due to the fact that a_U is the vertex of $(\Pi^{n-1})^\Delta$ corresponding to the facet of Π^{n-1} displayed in expression (3.3).

3.1.1 Pairwise absolute value mapping M

We will now introduce a mapping that will be central within these sections. We first state some basic properties and then consider the vertices of the permutahedron under this mapping.

$$M: \mathbb{R}^n \longrightarrow \mathcal{S}\mathbb{M}(n)$$
$$x \mapsto \left(|x_k - x_l|\right)_{\substack{k=1,\ldots,n \\ l=1,\ldots,n}}$$

It maps an element of $\mathbb{R}^n$ to a symmetric $n \times n$ matrix with zero-entries in the main diagonal. We start with some basic properties of M.

Lemma 3.1.8. *The mapping M has the following properties.*

(a) *For all $\xi \in \mathbb{R}$ and $x \in \mathbb{R}^n$, we have $M(x + \xi \mathbf{1}) = M(x)$.*

(b) *For $x, y \in L^n$, we have*

$$M(x) = M(y) \quad \text{if and only if} \quad x = y \text{ or } x = -y.$$

(c) *The mapping M is linear on each of the cones N_π, and it is also injective there.*

(d) *For each π, the image of N_π under M is a $(n-1)$-dimensional simplicial cone with apex 0 in $\mathcal{S}\mathbb{M}(n)$, which is generated by the extreme rays $\mathbb{R}_+ M(\chi^U)$ for all nonempty subsets $U \subsetneq [n]$ incident on π.*

Proof. (a). We have

$$M(x + \xi \mathbf{1}) = \left(|x_k + \xi - (x_l + \xi)|\right)_{\substack{k=1,\ldots,n \\ l=1,\ldots,n}} = \left(|x_k - x_l|\right)_{\substack{k=1,\ldots,n \\ l=1,\ldots,n}} = M(x).$$

(b). It is true that $x = \pm y \Rightarrow |x| = |y| \Rightarrow |x_k - x_l| = |y_k - y_l|$ for all $1 \le k, l \le n$. Therefore $M(x) = M(y)$. For the non-trivial direction, let $x, y \in L^n$ and define $x' := x - x_1 \mathbf{1}$, $y' := y - y_1 \mathbf{1}$. Note that $M(x') = M(x) = M(y) = M(y')$. We will show that $x' = y'$, which implies $x = y$ because $\sum_j x_j = 1 = \sum_j y_j$ as $x, y \in L^n$. Since

$$|x'_k| = |x'_k - x'_1| = M_{k,1}(x') = M_{k,1}(y') = |y'_k - y'_1| = |y'_k|,$$

3.1. Preparative Definitions

we conclude that the index set $[n]$ is the union of the three disjoint sets I_+, I_0, and I_- defined by $x'_k = y'_k \neq 0$ for all $k \in I_+$, $x'_k = y'_k = 0$ for all $k \in I_0$, and $x'_k = -y'_k \neq 0$ for all $k \in I_-$. We show that one of the sets I_+ or I_- must be empty. Assume the contrary: $k \in I_+$ and $l \in I_-$. Then $|x'_k - x'_l| = |y'_k - y'_l| = |x'_k + x'_l|$, hence $|x'_k - x'_l|^2 = |y'_k - y'_l|^2 = |x'_k + x'_l|^2$, which is equivalent to $x'^2_k - 2x'_k x'_l + x'^2_l = x'^2_k + 2x'_k x'_l + x'^2_l$. It follows $x'_k x'_l = 0$, thus we have $k \in I_0$ or $l \in I_0$, a contradiction.

(c). Let π be a permutation. We have to show that, for each k,l with $k \neq l$, the restriction of the mapping $x \mapsto M_{k,l}(x)$ to N_π is linear. Let such k,l be given. Then by Corollary 3.1.4, we are in one of the following two situations:

- $\pi(k) > \pi(l)$ and for any $x \in N_\pi$, we have $|x_k - x_l| = x_k - x_l$, or

- $\pi(k) < \pi(l)$ and for any $x \in N_\pi$, we have $|x_k - x_l| = -x_k + x_l$.

In both cases $x \mapsto M_{k,l}(x)$ is linear on N_π. Since $x \in \text{kernel}(M_{|N_\pi})$ if and only if $x_k = x_l$ for all $1 \leq k,l \leq n$, but by the definition of L^n exists exactly one $x \in L^n$ that fulfills this property: $x = \mathbf{0}$, the injectivity follows from Lemma 3.1.8-(b).

(d). By the previous items and Lemma 3.1.6, we know that $M(N_\pi)$ is the image of an $(n-1)$-dimensional simplicial cone with apex zero under an injective linear mapping. $\square$

As $M(x + \xi \mathbf{1}) = M(x)$ for all $\xi \in \mathbb{R}$ and $x \in \mathbb{R}^n$, we restrict M to the space orthogonal to $\mathbb{R}\mathbf{1} = \{\xi \mathbf{1} \mid \xi \in \mathbb{R}\}$ which is the linear vector space L^n which has already been mentioned above. We call the restriction of the mapping again M. The dual space $(L^n)^*$ is identified with L^n via the isomorphism $L^n \to (L^n)^* : c \mapsto \langle c |$, where $\langle c | : L^n \to \mathbb{R}$ is the mapping such that $y \mapsto \langle c | y \rangle$.

Remark 3.1.9. *Recall the definition of a permutation matrix E_π defined in Section 0.3.2 on page 5. The function $M \mapsto E_\pi^T M E_\pi$ is a linear isomorphism of the vector space $\mathsf{S}_n\mathsf{M}$ that maps P_n to P_n and Q_n to Q_n. Defining $(x \circ \sigma)_j := x_{\sigma(j)}$ for $x \in \mathbb{R}^n$ and $\sigma \in S(n)$, we obtain $E_\sigma^T M(x) E_\sigma = M(x \circ \sigma)$.*

We will now consider the vertices of the permutahedron under the mapping M. By Lemma 3.1.8-(a) and -(b) we have $M(\pi) = M(v_\pi) = M(-v_\pi) = M(\pi^-)$. As $M(\chi^U) = M(\chi^{U^c})$, we have that $M(\pi)$ corresponds to the pair U and U^c. Note that $M(\chi^U)$ is a cut matrix and, after appropriate permutation, looks like

$$M(\chi^U) = \begin{pmatrix} \mathbb{1} & \mathbb{0} \\ \mathbb{0} & \mathbb{1} \end{pmatrix} \begin{matrix} \}|U| \\ \}|U^C|, \end{matrix} \qquad M(\chi^U)^T = \begin{pmatrix} \mathbb{0} & \mathbb{1} \\ \mathbb{1} & \mathbb{0} \end{pmatrix} \begin{matrix} \}|U| \\ \}|U^c|, \end{matrix}$$

therefore we say **vertex $M(\pi)$ and cut (U, U^c) are incident** if and only if π and U as well as π^- and U^c are incident.

Note: as we have a bijection between sets the sets U and the facets (3.3), it does not matter if we consider $S(n)$ to be in L^n or in $\mathbb{R}^n$. The bijection is conserved, hence the direction $M(\chi^U)$ stays the same under translation.

For a permutation $\pi \in S(n)$, we call a set $U \subsetneq [n]$ **over the ridge from π** if

$$U = \{\pi^{-1}(1), \ldots, \pi^{-1}(k-1), \pi^{-1}(k+1)\} \quad \text{and}$$
$$U^c = \{\pi^{-1}(k), \pi^{-1}(k+2), \ldots, \pi^{-1}(n-1)\},$$

where $k = |U|$ [1]. We say $M(\chi^U)$ is over the ridge from $M(\pi)$ if U is over the ridge from π. Note: U is over the ridge from π if and only if U corresponds to a facet of the permutahedron which contains a neighbor vertex of π but not π itself.

---
[1] The terminology is taken from the geometry of the permutahedron: Since the polar polytope of the permutahedron, is simplicial, if we start somewhere "on π" and "walk over" a particular ridge to a neighboring facet π', then a unique vertex "comes into sight." The sets corresponding to these vertices are precisely those which are "over the ridge" from π.

3.2 Integral Polyhedron P_n

Having introduced these definitions and basic properties, we can start investigating the polyhedral aspects of the MinLA. We start with a short literature review about $P_n(G)$ and $\text{DOM}_n(G)$.

3.2.1 Literature Review

Definition of P_n and Basic Properties

The definition of the polyhedron corresponding to the y-variable formulation is

$$P_n(G) := \text{conv}\left\{y \in \mathbb{R}_+^{|E|} \;\middle|\; \text{exists } \pi \in S(n) \text{ such that } y_{ij} = |\pi(i) - \pi(j)| \text{ for all } ij \in E\right\}$$

and was given in [104]. Here, a class of polyhedra associated with the so-called Single Row Facility Layout Problem is investigated. As the MinLA is a special case, several polyhedral results are important for our case. It is shown that $P_n := P_n(K_n)$ is of dimension $\binom{n}{2} - 1$ and that its affine hull is defined by the equation $\sum_{i,j \in [n]} y_{ij} = \binom{n+1}{3}$. It is also shown that the following four classes of inequalities define facets of P_n under mild conditions:

- Pure hypermetric inequalities, which are simply the hypermetric inequalities from Section 1.5.2 for which $b \in \{0, \pm 1\}^n$;

- Strengthened pure negative-type inequalities, which are like the negative-type inequalities

$$\sum_{i,j \in [n]} b_i b_j y_{ij} \leq 0 \qquad \text{for all } b \in \mathbb{R}^n \text{ such that } \sum_{i=1}^n b_i = 0$$

  for which $b \in \{0, \pm 1\}^n$ except that the right-hand side is increased from 0 to $\frac{1}{2} \sum_{i \in [n]} |b_i|$;

- Clique inequalities, see Section 1.5.2;

- Strengthened star inequalities, which take the form

$$(|S| - 1) \sum_{i \in S} y_{ri} - \sum_{i,j \in S} y_{ij} \geq \lfloor (|S| + 1)^2 (|S| - 1)/12 \rfloor,$$

  where $r \in V$ and $S \subseteq V \setminus \{r\}$ with $|S| \geq 2$.

It is pointed out in the same paper that each star inequality with $|S| \geq 2$ is dominated by a clique inequality and a strengthened star inequality. Therefore, very few of the star inequalities define facets of P_n. Nevertheless, it is possible to strengthen the stars in such a way that these strengthened stars define facets of P_n. Furthermore, it is shown under which conditions odd wheels and 2-chorded cycle inequalities of different types are facet-defining for P_n.

In [104] the polytope $P_n(G)$ is investigated for general graphs G as well. In contrast to the polytope of the complete graph $P_n(G)$ is full dimensional if G is connected but not complete. Furthermore, a lot of facets can be "borrowed" from P_n. Cliques inequalities, pure hypermetric inequalities, strengthened pure negative-type inequalities, strengthened star inequalities, odd wheel inequalities, and 2-chorded cycle inequalities define facets of $P_n(G)$ whenever G contains a subgraph with a suitable structure. Besides that, the conditions are given under which the metric and star inequalities define facets of $P_n(G)$.

DOM_n and $\text{DOM}(P_n(G))$

As P_n has a fairly complex structure, a suitable relaxation was considered. As we consider a minimization problem with non-negative objective function coefficients, it is equivalent to optimize over P_n or over

$$\text{DOM}_n := P_n + \{y \in \mathbb{R}^{|E|}+ \mid \text{exists } y' \in P_n \text{ such that } y' \geq y\}.$$

3.2. Integral Polyhedron P_n

DOM_n is called the **dominant** of P_n. In [7] it is shown that DOM_n is full dimensional and unbounded. Furthermore, P_n is the unique bounded facet of DOM_n. Several results are given that the following inequalities are facet-defining under mild conditions:

- star inequalities,
- clique inequalities,
- circuit inequalities,
- bipartite inequalities, and
- double star inequalities.

3.2.2 Feasibility Test for $P_n(G)$

We are faced with the problem of a graph with integral edge lengths and have to determine an embedding in $\mathbb{R}$ that fulfills all given edge lengths if it exists. We are going to characterize a graph property for which this problem can be solved efficiently with the presented algorithm.

Concept

The key idea is that a nodes' position in a permutation can exactly be given if it has two adjacent nodes that are already positioned within the permutation. We call $u_1, \ldots, u_n$ a k-**convenient node ordering** if $u_1, \ldots, u_k$ are connected and node u_{j+1} has two neighbors in $u_1, \ldots, u_j$ for all $j \geq k$. A graph is k-**spanning** if a k-convenient node ordering of the nodes of G exists.

Algorithm

We use the property described above in an embedability test displayed in Algorithm 5. It starts with the test whether

---

Algorithm 5 `Embedability Test`

Input: Connected graph $G = (V, E)$ with edge lengths y,
Constant k,
Output: Mapping $V \to \mathbb{R}$ which is in correspondence to y.

1: Determine whether G has a k-convenient node ordering, see Algorithm 6.
2: **if** G is k-spanning **then**
3:    Take a k-convenient node ordering $u_1, \ldots, u_n$ of G.
4: **else**
5:    Choose any ordering of the nodes $u_1, \ldots, u_n$ such that u_j has at least one neighbor in $u_1, \ldots, u_{j-1}$ for all $2 \leq j \leq n$. // As G is connected such an ordering can iteratively be constructed.
6: **end if**
7: Determine all possible embeddings of $u_1, \ldots, u_n$, see Algorithm 7.

---

a k-convenient ordering exists. The complete procedure is realized in Algorithm 6 and analyzed in Proposition 3.2.1.

Proposition 3.2.1. *For fixed k, Algorithm 6 determines a k-convenient node ordering $u_1, \ldots, u_n$ of G if such an ordering exists.*

Algorithm 6 `Determine if G is k-spanning`

Input: Connected graph $G = (V,E)$ with edge lengths y,
Output: Node ordering $u_1,\ldots,u_n$, if possible k-convenient.

1: **for all** $X \subset V, |X| = k$, induced subgraph is connected **do**
2:    Orientate all edges in $\delta(X)$ to point away from X. No other edge is oriented.
3:    $Y := X = \{u_1,\ldots,u_k\}$.
4:    **for all** $j = k+1,\ldots,n$ **do**
5:      **if** no $u \in V \setminus Y$ exists with $\deg_{in}(u) \geq 2$ **then**
6:        Go to step 12.
7:      **end if**
8:      Let u_j be any node in $V \setminus Y$ with $\deg_{in}(u_j) \geq 2$.
9:      Orientate those edges in $\delta(u_j)$ that have not already been oriented to point away from u_j.
10:      Set $Y := Y \cup \{u_j\}$.
11:    **end for**
12:    **if** $Y = V$ **then**
13:      Return "k-convenient node ordering $u_1,\ldots,u_n$".
14:    **end if**
15: **end for**
16: Return "G is not k-spanning".

Algorithm 7 `Determine all possible embedings of G`

Input: Ordering $u_1,\ldots,u_n$ of nodes with edge lengths $y_{|\{u_1,\ldots,u_n\}}$,
          Constant k,
Output: Possibly empty list Π_n of all injective embeddings p of G.

1: Find all injective embeddings $p : \{1,\ldots,k\} \to \mathbb{R}$ of $y_{|\{u_1,\ldots,u_k\}}$, Algorithm 8.
2: Save these embedings in the list Π_k.
3: **for all** $j = k+1,\ldots,n$ **do**
4:    **for all** embedings p in Π_{j-1} **do**
5:      **if** p can be continued to p' of $y_{|\{u_1,\ldots,u_j\}}$, Algorithm 9 **then**
6:        Continue p to p'.
7:        Save p' in list Π_j of embedings of $y_{|\{u_1,\ldots,u_j\}}$.
8:      **end if**
9:    **end for**
10: **end for**
11: Return possibly empty list Π_n of all injective embedings p of G.

3.2. INTEGRAL POLYHEDRON P_n

Algorithm 8 `Determine all possible embeddings of` $u_1,\ldots,u_k$

Input: Ordering $u_1,\ldots,u_k$ of nodes with edge lengths $y_{|\{u_1,\ldots,u_k\}}$,
Output: Possibly empty list Π_k of all injective embeddings p of $y_{|\{u_1,\ldots,u_k\}}$.

1: Determine the injective embedding $p : \{1,2\} \to \mathbb{R}$ of $y_{|\{u_1,u_2\}}$, where w.l.o.g. $p(u_1) := 0$ and $p(u_1) < p(u_2)$.
2: Save p in the list Π_2.
3: **for all** $j = 3,\ldots,k$ **do**
4:    **for all** embeddings p in Π_{j-1} **do**
5:       **if** p can be continued to p' of $y_{|\{u_1,\ldots,u_j\}}$, Algorithm 9 **then**
6:          Continue p to p'.
7:          Save p' in list Π_j of embeddings of $y_{|\{u_1,\ldots,u_j\}}$.
8:       **end if**
9:    **end for**
10: **end for**
11: Return possibly empty list Π_k.

Algorithm 9 `Check existence of continuation`

Input: Nodes $u_1,\ldots,u_{k-1}$ and u_k,
         Embedding $p \in \Pi_{k-1}$,
Output: Possibly empty list $(\Pi_k)_{|p_{k-1}}$ of all injective embeddings p of $y_{|\{u_1,\ldots,u_k\}}$ that
         are continuations of p_{k-1}.

1: Choose node $u_l \in \{u_1,\ldots,u_{k-1}\}$ adjacent to u_k. // Exists due to the chosen node ordering.
2: Use distance $y_{u_l u_k}$ to compute the two possible continuations p_1, p_2 of p that place node u_k. // Note: both embeddings are injective
3: Let *mistake* be a bool variable.
4: **for all** embeddings p_i, $i = 1, 2$ **do**
5:    Set *mistake*=False.
6:    **for all** $j = 1,\ldots,k-1$ with $j \neq l$ **do**
7:       **if** $y_{u_j u_k} \neq |p_i(u_j) - p_i(u_k)|$ **then**
8:          Set *mistake*=True.
9:       **end if**
10:    **end for**
11:    **if** *mistake*=False **then**
12:       Save injective embedding p_i in $(\Pi_k)_{|p_{k-1}}$.
13:    **end if**
14: **end for**
15: Return possibly empty list $(\Pi_k)_{|p_{k-1}}$.

Proof. Assume there exists a k-convenient node ordering of G. We will prove that a set X exists for which the following invariant holds: At the beginning of the loop in line 4 of Algorithm 6, there exists a k-convenient node ordering starting with nodes $u_1, \ldots, u_{j-1}$. We construct the set X and show that the invariant is not destroyed within any iteration of the loop.

Let X consist of the first k nodes from the existing k-convenient node ordering $u_1, \ldots, u_n$ of G, i.e., $X := \{u_1, \ldots, u_k\}$. It is clear that the invariant holds for the beginning $j = k+1$. Therefore, at least one node $v \in \{u_j, \ldots, u_n\}$ exists with $\deg_{in}(v) \geq 2$. The algorithm chooses one of these nodes v and moves it to position j, i.e., the algorithm sets $u_j := v$. Note: node v is moved to the left side—if it is moved at all. It therefore remains to be a left-hand-side neighbor to all its right-hand-side-neighbors. This means that we do not loose the k'-convenient property for any $k' > k$ changing the position of v. Hence, at the beginning of the loop iteration $j+1$ there exists a k-convenient node ordering starting with the nodes $u_1, \ldots, u_{j-1}, u_j$. □

If the graph is k-spanning, we continue with its k-convenient ordering and obtain a polynomial running time.

Proposition 3.2.2. *For fixed k, if $u_1, \ldots, u_n$ is a k-convenient node ordering of G, Algorithm 7 is polynomial in time and space.*

Proof. For a k-convenient node ordering, the position of a node within an embedding is unique if its two neighbors have already been placed. Hence, the exponential growth of the list Π_i is stopped when $i \geq k$. The total number of possible embeddings is therefore 2^k. This leads to a linear worst case running time and worst case memory for Algorithm 7. □

Although this running time cannot be guaranteed for arbitrary graphs, the feasibility test holds for all kinds of graphs, as it is shown in the following proposition.

Proposition 3.2.3. *Algorithm 5 is correct, i. e., it constructs all feasible embeddings corresponding to the given distances.*

Proof. We show that every feasible embedding is found by Algorithm 5. Consider an arbitrary feasible embedding p. We examine two situations.

Case 1: G is k-spanning. Let $u_1, \ldots, u_n$ be a k-convenient node ordering and let $p_j \in \Pi_k$ be the embedding of $u_1, \ldots, u_k$ with $p(u_i) = p_j(u_i)$ for all $i = 1, \ldots, k$. In every iteration of step 3 in Algorithm 7 the embeddings are continued to embed one more node. As $u_1, \ldots, u_n$ is k-convenient, every newly placed node has two neighbors. Hence, its position is unique and therefore the continuation of p_j is uniquely determined in each iteration. Hence, in the last iteration p_{n-1} is continued to the embedding p.

Case 2: G is not k-spanning. Consider any node ordering $u_1, \ldots, u_n$ in which u_i has at least one neighbor in $\{u_1, \ldots, u_{i-1}\}$ for all $i = 2, \ldots, n$. Again let $p_j \in \Pi_k$ be the embedding of $u_1, \ldots, u_k$ with $p(u_i) = p_j(u_i)$ for all $i = 1, \ldots, k$. In every iteration, at most two continuations of p_j are constructed which leads to at most 2^{n-k} embeddings of $\{u_1, \ldots, u_n\}$, all being continuations of p_j. As the algorithm constructs every possible continuation in each iteration, p is one of the 2^{n-k} embeddings. □

We will now state some open questions for further research.

- The complexity of the embedability problem in $\mathbb{R}$ is known to be $\mathcal{NP}$-hard for semi-metrics. For metrics, i.e., for injective embeddings, the complexity is not known. What is the complexity of the injective embedability problem for arbitrary graphs?

- Although trees can trivially be embeded for semi-metrics, it is not even known how complex their embedding is for metrics. What is the complexity of the injective embedability problem for trees?

- Can the k-spanning graphs be characterized?

    - G has to be bi-connected and its circumference has to be $\leq k-1$. Unfortunately, these conditions are not sufficient.

- Which reasonable, "nice" graph property implies k-spanningness?
- Consider non bi-connected graphs. Can a possibly polynomial algorithm for injective embedding of trees be modified to give a polynomial embedding algorithm for graphs all of whose blocks are k-spanning for a fixed k?
- If G is not k-spanning, but, say, there is a long path between two vertices u_{l_1}, u_{l_2}, none of whose vertices has a neighbor in $u_1, \ldots, u_j$. Can we use a modification of the dynamic programming algorithm for knapsack problems to embed such a graph?

3.3 Convex Set Q_n

We now turn our attention to Q_n and $\overline{Q_n}$, which is an alternative relaxation of the polytope P_n. In contrast to DOM_n, we do not consider the Minkowski sum of P_n and the non-negative orthant of $\mathbb{R}^{\binom{n}{2}}$ but the Minkowski sum of P_n and the cut cone CUT_n. We prove several results about the structure of this convex set Q_n. In particular, it is shown that it is not closed in general. We characterize some of the $(n-1)$-dimensional faces (i.e., facets) of the closure and some of the 1-dimensional faces (i.e., edges) of both the convex hull and its closure.

3.3.1 Definition of Q_n and Basic Properties

We consider the following convex set:

$$Q_n := \mathrm{conv}\left\{ y \in \mathbb{R}_+^{\binom{n}{2}} \,\middle|\, \text{exists } z \in \mathbb{R}^n \text{ such that } y_{ij} = |z_i - z_j| \geq 1 \text{ for all } i \neq j \in [n] \right\}.$$

To give some intuition, we present in Figure 3.1 drawings of Q_3 from three different angles. Of course, the drawing is truncated, since Q_3 is unbounded. The three co-ordinates represent y_{12}, y_{13} and y_{23}. The three coloured regions represent the three disjoint subsets of Q_n that we will characterize in Theorem 3.3.1. One can see that Q_3 is a three-dimensional polyhedron with one bounded facet, six unbounded facets, three bounded edges and six unbounded edges. To get an appropriate description of Q_n we will now give a structural theorem about Q_n relating it to the normal fan of the permutahedron. The use of the pairwise absolute value mapping M will be central within the proof of the theorem.

Theorem 3.3.1. *The set Q_n is the convex hull of $(n!/2)$ pairwise disjoint $(n-1)$-dimensional simplicial cones of the form $M(\pi) + M(N_\pi)$, i.e.,*

$$Q_n = \mathrm{conv}\left\{ \bigcup_{\pi \in S(n)} M(\pi) + M(N_\pi) \right\}.$$

Two cones $M(\pi) + M(N_\pi)$ and $M(\pi') + M(N_{\pi'})$ are identical if π' and π are identical or antipodal; otherwise they are disjoint.

Proof. From the definition of Q_n, we immediately get by Lemma 3.1.8-(a) that

$$Q_n = \mathrm{conv}\left\{ M\left(\{x \in L^n \mid |x_k - x_l| \geq 1 \text{ for all } 1 \leq k \neq l \leq n\} \right) \right\}.$$

As the normal fan is complete, it follows that

$$M\left(\{x \in L^n \mid |x_k - x_l| \geq 1 \text{ for all } 1 \leq k \neq l \leq n\} \right)$$
$$= \bigcup_{\pi \in S(n)} M\left(\{x \in N_\pi \mid |x_k - x_l| \geq 1 \text{ for all } 1 \leq k \neq l \leq n\} \right).$$

In Proposition 3.1.7 we have proven $\{x \in N_\pi \mid |x_k - x_l| \geq 1 \text{ for all } 1 \leq k \neq l \leq n\} = v_\pi + N_\pi$. And as M is linear and injective on each N_π and $M(\pi) = M(v_\pi)$, by Lemma 3.1.8-(c) we conclude

$$M\left(\{x \in N_\pi \mid |x_k - x_l| \geq 1 \text{ for all } 1 \leq k \neq l \leq n\} \right)$$
$$= M(v_\pi + N_\pi) = M(v_\pi) + M(N_\pi) = M(\pi) + M(N_\pi).$$

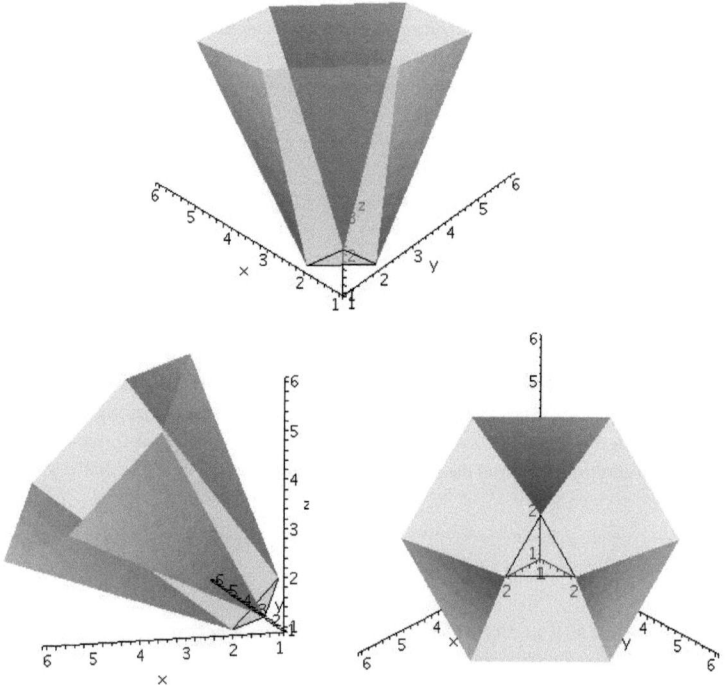

Figure 3.1: Convex set Q_3 in $M(L^3)$ dispayed from three perspectives.

3.3. The Convex Set Q_n

Putting all arguments together, we obtain

$$Q_n = \text{conv}\Big\{M\big(\{x \in L^n \mid |x_k - x_l| \geq 1 \text{ for all } 1 \leq k \neq l \leq n\}\big)\Big\}$$
$$= \text{conv}\Big\{\bigcup_{\pi \in S(n)} M\big(\{x \in N_\pi \mid |x_k - x_l| \geq 1 \text{ for all } 1 \leq k \neq l \leq n\}\big)\Big\}$$
$$= \text{conv}\Big\{\bigcup_{\pi \in S(n)} M\big(v_\pi + N_\pi\big)\Big\}$$
$$= \text{conv}\Big\{\bigcup_{\pi \in S(n)} M(\pi) + M(N_\pi)\Big\}.$$

The cones $M(N_\pi)$ are simplicial by Lemma 3.1.8-(d).

For the last part of the proof, consider

$$M(v_\pi + N_\pi) = M(-(v_\pi + N_\pi)) = M(v_{\pi^-} + N_{\pi^-}).$$

Hence the maximal number of distinct cones is $n!/2$. From the outer description of the cones N_π, see Lemma 3.1.4, follows immediately that all sets $\{x \in N_\pi \mid |x_k - x_l| \geq 1 \text{ for all } 1 \leq k \neq l \leq n\}$ are pairwise disjoint. The intersection of two such sets with π and π' can only be nonempty if $\pi' = \pi^-, \pi$. Hence, because of Lemma 3.1.8-(b) it follows that the cones $M(\pi) + M(N_\pi)$ and $M(\pi') + M(N_{\pi'})$ are equal if and only if π and π' are identical or antipodal. The cones are disjoint in any other case. Therefore we have $n!/2$ pairwise disjoint cones. □

We note some consequences of the theorem.

Lemma 3.3.2.

(a) Q_n is the convex hull of all the half-lines $M(\pi) + \mathbb{R}_+ M(\chi^U)$ where π is a permutation of $[n]$ and U is a nonempty proper subset of $[n]$ such that π and U are incident.

(b) Q_n is a full-dimensional unbounded convex set.

(c) Q_n contains P_n as an exposed subset: the inequality $\mathbb{1}_n \bullet X \geq 2\binom{n+1}{3}$ is valid for Q_n.

(d) The extreme points of Q_n are precisely the vertices of P_n which are the of the form $M(\pi)$ for $\pi \in S(n)$.

Proof. (a) These are exactly the extreme rays of the cones whose convex hull is equal to Q_n.

(b) As P_n is contained in Q_n but the only valid equation for P_n is not valid for Q_n, Q_n has to be full-dimensional.

(c) and (d) Follow from (a). □

3.3.2 Unbounded Edges of Q_n

We now want to investigate *how* the cones $M(\pi) + M(N_\pi)$ are subsets of Q_n. Considering Figure 3.1, it can be seen that in the case $n = 3$, the three cones are faces of Q_3 (as Q_3 is closed, we can safely speak of faces). In the following we show that this is the case for all n, and we also characterize the extremal half-lines of Q_n. This will be useful in comparing Q_n with its closure: We will characterize the unbounded edges issuing from each vertex for the polyhedron $\overline{Q_n} = P_n + \text{CUT}_n$ in the following subsection.

We are dealing with an unbounded convex set of which we do not know whether it is closed or not. In fact, we will show that Q_n is almost never closed. For this purpose, we supply the following two lemmas for easy reference.

Lemma 3.3.3. *Let S be a set, $x \in S$, and $y \neq 0$ such that $x + \mathbb{R}_+ y$ is an extreme subset of $\text{conv}(S)$. Then*

$$\text{for all } \lambda \in \mathbb{R}_+ \text{ exists } \mu \geq \lambda : \ x + \mu y \in S.$$

Proof. Let $m = |S|$. For any $\lambda \in \mathbb{R}_+$, we have $x + \lambda y \in \text{conv}(S)$, hence $x + \lambda y = \sum_{j=1}^{m} t_j s_j$, for $\mathbf{1}^T t = 1$, $t_j \geq 0$ and $s_j \in S$ for all $1 \leq j \leq n$. As $x + \lambda y$ is an extreme subset for all j where $t_j > 0$, we have $s_j = x + \mu_j y$ with $\mu_j \in \mathbb{R}_+$. This means

$$x + \lambda y = \sum_{j=1}^{m} t_j s_j = \sum_{j=1}^{m} t_j(x + \mu_j y)$$
$$= \big(\sum_{j=1}^{m} t_j\big) x + \big(\sum_{j=1}^{m} t_j \mu_j\big) y = x + \big(\sum_{j=1}^{m} t_j \mu_j\big) y.$$

This is equivalent to $0 = (\sum_{j=1}^{m} t_j \mu_j - \lambda)y$, and as $y \neq 0$ this implies $\sum_{j=1}^{m} t_j \mu_j = \lambda$. That means there exists $1 \leq k \leq m$ such that $\mu_k \geq \lambda$ and $x + \mu_k y = s_k \in S$. □

Lemma 3.3.4. *For $k = 1, \ldots, m$ let K_k be a (closed) polyhedral cone with apex x_k. Suppose that the K_k are pairwise disjoint and define $S := \bigcup_{k=1}^{m} K_k$. Let x, y be vectors such that $x + \mathbb{R}_+ y$ is an extreme subset of $\mathrm{conv}(S)$. The following holds:*

(a) *There exists $\lambda_0 \in \mathbb{R}_+$ and a k such that $x + \lambda y \in K_k$ for all $\lambda \geq \lambda_0$.*

(b) *There exists $\lambda_1 \in \mathbb{R}_+$ and a k such that $x_k = x + \lambda_1 y$ and $x_k + \mathbb{R}_+ y = \{x + \lambda y \mid \lambda \geq \lambda_1\}$ is an extreme ray of the polyhedral cone K_k.*

Proof. (a). There are only finitely many cones K_k. Therefore, by the previous lemma, there will be a point $x + \lambda_0 y$ from which on we stay within a certain cone K_k. Hence, $x + \lambda y \in K_k$ for all $\lambda \geq \lambda_0$.

(b). By the previous item we know $x + \lambda y \in K_k$ for all $\lambda \geq \lambda_0$. On the other hand $x + \mathbb{R}_+ y$ is extreme and therefore cannot be the conic combination of other extreme subsets. Hence $\{x + \lambda y \mid \lambda \geq \lambda_0\}$, which is the part of $x + \mathbb{R}_+ y$ laying in K_k, has to be an extreme ray of the polyhedral cone K_k. Note that λ_1 is the smallest of all λ_0 of (a). □

Proposition 3.3.5.

(a) *For every $\pi \in S(n)$, each face of the cone $M(\pi) + M(N_\pi)$ is an exposed subset of Q_n.*

(b) *The unbounded one-dimensional extremal sets of Q_n are exactly the defining half-lines. In other words, every half-line $X + \mathbb{R}_+ Y$ which is an extremal subset of Q_n is of the form $M(\pi) + \mathbb{R}_+ M(\chi^U)$ for a $\pi \in S(n)$ and a set U incident to π. In particular, for every vertex $M(\pi)$ of Q_n, the unbounded one-dimensional extremal subsets of Q_n containing $M(\pi)$ are in bijection with the nonempty proper subsets of $[n]$ incident to π. Thus, there are precisely $n - 1$ of them.*

Proof. (a). Due to Remark 3.1.9, it is sufficient to treat the case $\pi = 1 := (1, \ldots, n)^T$, the identity permutation. Consider the matrix

$$C := \begin{pmatrix} 0 & 1 & & & -1 \\ 1 & 0 & 1 & & 0 \\ & 1 & \ddots & & \\ & & & & 1 \\ 0 & & & 1 & 0 & 1 \\ -1 & & & & 1 & 0 \end{pmatrix} \in S_0 \mathbb{M}(n).$$

It is easy to see that the minimum over all $C \bullet M(\pi)$, $\pi \in S(n)$, is attained only in the case $\pi = 1, 1^-$ with the value 0. Moreover, for any nonempty proper subset U of $[n]$, we have $C \bullet M(\chi^U) = 0$ if U is incident to 1 and $C \bullet M(\chi^U) > 0$ otherwise. Hence, we have that $D(1) + M(N_1)$ is equal to the set of all points in Q_n which satisfy the valid inequality $C \bullet X \geq 0$ with equality. Out of this matrix C we will now construct a matrix C' and a right hand side such that only some of the subsets incident to 1 fulfill the inequality with equality. To do so, let U_0 be a subsets of $[n]$ incident to 1. If, for each $U \subset [n]$ incident to 1 but different from U_0, we increase the matrix entries $C_{\max U, \max U+1}$ and $C_{\max U+1, \max U}$ by one, we obtain an inequality $C' \bullet X \geq 0$ which is valid for Q_n and such that the set of all points of Q_n which are satisfied with equality is precisely the edge of $M(1) + M(N_1)$ generated by the half-lines $M(1) + \mathbb{R}_+ M(\chi^{U_0})$.

(b). That the defining half-lines are extremal has just been proven in (a). The converse statement follows from Lemma 3.3.4 and the fact that the extreme points of Q_n are precisely the vertices of P_n which are of the form $M(\pi)$ for $\pi \in S(n)$. □

3.3.3 The Minkowski Sum $P_n + \mathrm{CUT}_n$ equals $\overline{Q_n}$

Amaral and Letchford showed in [105] the following result.

Lemma 3.3.6. $Q_n \subseteq P_n + \mathrm{CUT}_n$.

We will therefore now consider the Minkowski sum of the cut cone and P_n. It is clear that P_n is a subset of $P_n + \mathrm{CUT}_n$, in fact we get

3.3. The Convex Set Q_n

Lemma 3.3.7. P_n is the only bounded facet of $P_n + \text{CUT}_n$.

Proof. We first prove that the equation $\mathbb{1} \bullet D = 2\binom{n+1}{3}$ is valid for $P_n + \text{CUT}_n$. As the affine hull of P_n is defined by this equation, it has only to be shown that $\mathbb{1} \bullet D \geq 0$ for all $D \in \text{CUT}_n$. For **case** 1, consider only the cut matrices D_S where $S \subseteq [n]$. As

$$D_S = \begin{pmatrix} \mathbb{1} & 0 \\ 0 & \mathbb{1} \end{pmatrix} \begin{matrix} \} \ |S| \\ \} \ n-|S|, \end{matrix} \qquad D_S^T = \begin{pmatrix} 0 & \mathbb{1} \\ \mathbb{1} & 0 \end{pmatrix} \begin{matrix} \} \ |S| \\ \} \ n-|S|, \end{matrix}$$

it follows $\mathbb{1} \bullet D_S = (n-|S|)|S| + |S|(n-|S|) \geq 0$. For **case** 2 consider any $D \in \text{CUT}_n$. It is true that $\mathbb{1} \bullet D = \mathbb{1} \bullet \left(\sum_{S \in [n]} \lambda_S D_S \right) = \sum_{S \in [n]} \lambda_S (\mathbb{1} \bullet D_S)$, where $\lambda_S \geq 0$ for all $S \in [n]$. From the first case we know $\mathbb{1} \bullet D_S \geq 0$ for all S and therefore $\mathbb{1} \bullet D \geq 0$ is proven for all $D \in \text{CUT}_n$.

The dimension of a Minkowski sum is at least the maximal dimension of its polyhedra. As the cut cone CUT_n is full-dimensional, we immediately obtain that the dimension of $P_n + \text{CUT}_n$ is $\binom{n}{2}$. As the dimension of P_n is $\binom{n}{2} - 1$, we get that P_n is indeed a facet of $P_n + \text{CUT}_n$.

Now let F be a bounded facet of $P_n + \text{CUT}_n$. As F is bounded, it does not contain any extreme rays, hence F is contained in P_n and $\text{vert}(F) \subseteq \text{vert}(P_n)$. As facets are maximal, this implies $F = P_n$. □

We now prove that the closure of Q_n is equal to $P_n + \text{CUT}_n$.

Proposition 3.3.8. *The closure of Q_n is equal to the Minkowski sum $P_n + \text{CUT}_n$.*

Proof. From Lemma 3.3.6 we know that $P_n + \text{CUT}_n \subset \overline{Q_n}$ is impossible. Therefore assume $\overline{Q_n} \not\subset P_n + \text{CUT}_n$ and let x be an element of $P_n + \text{CUT}_n \setminus \overline{Q_n}$. Define $A \coloneqq \overline{Q_n}$ and $A' \coloneqq \{x\}$, then A, A' are convex and A' is bounded. We can apply the separating hyperplane lemma of [106] which says that there exists a hyperplane that separates A and A' strictly. W.l.o.g. $A \subset H^+$ and $A' \subset H^-$. As $x \in P_n + \text{CUT}_n$, we get $x = p + \lambda r$, with $p \in P_n, r \in \text{CUT}_n$ and $\lambda \in \mathbb{R}_+$. It is clear that $\lambda \neq 0$ and r is not $\mathbf{0}$. There are two cases to distinguish. In **case** 1, $\pi \in S(n)$ exists with $p = M(\pi)$ and $r \in M(N_\pi)$, hence $x \in M(\pi) + M(N_\pi) \subseteq Q_n$. As this contradicts the existence of H, this case cannot occur.

In **case** 2, different $\pi, \pi' \in S(n)$ exist with $p = M(\pi)$ and $r \in M(N_{\pi'})$. Now transform the coordinates such that $p = (0,1) \in \mathbb{R}^2$ and $r = (1,0) \times \mathbb{R}_+$, hence $x = (0,1) + \lambda(1,0)$ for $\lambda \in \mathbb{R}_+$. Then $x \notin \mathbb{R}_+ \times [0,1[= \text{conv}\{(0,1), \mathbb{R}_+ \times (1,0)\}$, but x is an element of the closure of the convex hull, $x \in \mathbb{R}_+ \times [0,1] = \overline{\text{conv}}\{(0,1), \mathbb{R}_+ \times (1,0)\}$. On the other hand, $\text{conv}\{(0,1), \mathbb{R}_+ \times (1,0)\} = \text{conv}\{M(\pi), M(N_{\pi'})\} \subset Q_n \subset H^+$ and as H is closed, we have $\overline{\text{conv}}\{(0,1), \mathbb{R}_+ \times (1,0)\} \subset H^+$. Again this is a contradiction, hence $\overline{Q_n} = P_n + \text{CUT}_n$. □

We can now add a note about the first part of Proposition 3.3.5.

Remark 3.3.9. *In the proof of part (a) of Proposition 3.3.5 we have actually proven that for every set $\{U_1, \ldots, U_r\}$ of nonempty proper subsets of $[n]$ incident to π, there is a matrix C such that the minimum $C \bullet M(\sigma)$ over all $\sigma \in S(n)$ is attained only in π and π^-, and that $C \bullet M(U') \geq 0$ for every nonempty proper subset of $[n]$ where equality holds precisely for the sets U_i and their complements. Knowing that $\overline{Q_n} = P_n + \text{CUT}_n$, by Proposition 3.3.8, our proof implies that $M(\pi) + \text{cone}\{M(\chi^{U_1}), \ldots, M(\chi^{U_r})\}$ is a face of the polyhedron $\overline{Q_n}$.*

Amaral and Letchford considered in [45] how a facet of the cut cone CUT_n can be transformed into a facet of the Minkowski sum $P_n + \text{CUT}_n$. To do so, let $ax \geq 0$ be facet defining for CUT_n and let x^* be an extreme point of P_n such that ax^* is minimum. Then the inequality $ax \geq ax^*$ is the so-called **cut cone facet** of $P_n + \text{CUT}_n$. It can easily be shown that this yields a facet of the Minkowski sum $P_n + \text{CUT}_n$ [45].

This raises the question whether cut cone facets of $P_n + \text{CUT}_n$ are facets of P_n. Therefore, we looked at the polytope P_n and the cut cone CUT_n more closely for $n = 5, 6$ and 7. Given a cut cone facet F, we denote by $\tilde{F}$ the face of P_n spanned by the facet defining inequality of F. We took all cut cone facets from the SMAPO library [107] and tested the dimension of $\tilde{F}$ and how many vertices of P_n are contained in $\tilde{F}$. The results presented in Table 3.1 show that $\tilde{F}$ can be a facet of P_n (see for $n = 5$ the second SMAPO facet, for $n = 6$ the second one or for $n = 7$ the SMAPO facets 5, 10 and 11) but generally this is not the case.

Table 3.1: Are there cut cone facets that are facets of P_n?

n	SMAPO	$|\text{vert}(\tilde{F})|$	$\dim(\tilde{F})$	$\dim(P_n)$
5	1	6	4	9
5	2	20	8	9
6	1	48	12	14
6	2	36	12	14
6	3	120	13	14
7	1	28	14	20
7	2	36	12	20
7	3	120	17	20
7	4	2	1	20
7	5	288	19	20
7	6	4	2	20
7	8	40	12	20
7	8	33	14	20
7	9	72	13	20
7	10	252	19	20
7	11	840	19	20

3.3. The Convex Set Q_n

Lemma 3.3.10. *Every cut cone facet F of $\overline{Q_n}$ has at least $\binom{n}{2} - 1$ extreme rays.*

Proof. From the definition of a cut cone facet F, see Section 3.3.3, there exists another facet F' which is facet defining for CUT_n. It is clear that $\dim(F') = \dim(\mathrm{CUT}_n) - 1 = \binom{n}{2} - 1$. As there is only one vertex in the cut cone, at least $\binom{n}{2} - 1$ extreme rays of CUT_n are necessary to achieve this dimension, hence it is clear that $|\mathrm{exray}(F')| \geq \binom{n}{2} - 1$.

As the cut cone facet F is by construction F' moved to a vertex in P_n, every extreme ray of F' is carried forward to an extreme ray of F. That means $|\mathrm{exray}(F)| \geq \binom{n}{2} - 1$. □

Lemma 3.3.11. *For every extreme ray $\mathbb{R}_+ r$ of $\overline{Q_n}$, there exists a vertex $v \in \overline{Q_n}$ such that $v + \mathbb{R}_+ r$ is a one-dimensional face of $\overline{Q_n}$.*

Proof. Let r be an extreme ray of Q_n and (a, b) the equality defined by r. As r has to start somewhere in Q_n, we know that $\min_{x \in P_n} ax$ is nonempty. In fact, $\min_{x \in P_n} ax = \{v \in L^n \mid v \in \mathrm{vert}(Q_n)\}$ as we optimize over a convex set and therefore the optima lie in the vertices of Q_n. That there exists a one dimensional face $v + \mathbb{R}_+ r$ follows from the fact that every unbounded polyhedron has endless edges. □

Proposition 3.3.12. *Let F be a cut cone facet of $\overline{Q_n}$. Then F has at least $\lceil \frac{n}{2} - \frac{1}{n-1} \rceil$ vertices.*

Proof. Let F be an arbitrary cut cone facet of Q_n. Further define a mapping

$$\phi : \mathrm{exray}(\mathrm{rec}(F)) \to \mathrm{vert}(F)$$

that maps an extreme ray r of the recession cone of F to the vertex $v \in Q_n$ where r starts. This vertex exists because of Lemma 3.3.11, hence

$$\mathrm{exray}(\mathrm{rec}(F)) = \dot{\bigcup}_{v \in \mathrm{vert}(F)} \{r \mid \phi(r) = v\}. \tag{3.4}$$

By Proposition 3.3.5 we know that every vertex of $\overline{Q_n}$ is contained in exact $n - 1$ edges, hence

$$\left|\{r \in \mathrm{exray}(\mathrm{rec}(F)) \mid \phi(r) = v\}\right| \leq n - 1 \quad \text{for all } v \in \mathrm{vert}(F).$$

Together with equation (3.4) this implies

$$|\mathrm{exray}(\mathrm{rec}(F))| = \left|\dot{\bigcup}_{v \in \mathrm{vert}(F)} \{r \mid \phi(r) = v\}\right|$$
$$\leq \sum_{v \in \mathrm{vert}(F)} (n - 1)$$
$$= |\mathrm{vert}(F)|(n - 1).$$

From Lemma 3.3.10 we have $\binom{n}{2} - 1 \leq |\mathrm{exray}(F)| = |\mathrm{exray}(\mathrm{rec}(F))|$. Therefore we conclude $\binom{n}{2} - 1 \leq |\mathrm{vert}(F)|(n-1)$ which is equivalent to $|\mathrm{vert}(F)| \geq \frac{n}{2} - \frac{1}{n+1}$ and this value can be rounded up because of the integrality. □

Do these results remain to be true for Q_n not closed in general? To answer this question, we consider Table 3.1. The cut cone facet F of $P_7 + C_7$ with SMAPO number 4 has $\mathrm{vert}(F) = \mathrm{vert}(\tilde{F}) = 2 \not\geq \lceil \frac{7}{2} - \frac{1}{7-1} \rceil = 4$. For this reason we know that cut cone facets in Q_n with $n \geq 4$ possibly contain only few vertices. In particular, we have another proof that $Q_7 \neq P_7 + C_7$.

Unbounded Edges of $\overline{Q_n}$

In the last section we already identified some unbounded edges of $\overline{Q_n}$ starting at a certain vertex $M(\pi)$ of $\overline{Q_n}$ (see Remark 3.3.9). We now want to characterize all unbounded edges of this polytope. From the definition of $\overline{Q_n}$ it is clear that all unbounded edges have the form $M(\pi) + \mathbb{R}_+ M(\chi^U)$, but we will see that not all of them are edges.

In the following we will say that $M(\pi) + \mathbb{R}_+ M(N_\pi)$ is the half-line **defined by the pair** $\pi \nearrow U$, where $\pi \in S(n)$ is a permutation and U is a nonempty, proper subset of $[n]$. We will now characterize the distinct pairs $\pi \nearrow U$ whose defining half-lines are edges of $\overline{Q_n}$.

60 CHAPTER 3. POLYHEDRAL THEORY

To do so, we introduce a more "visual" notation of the sets $U \subsetneq [n]$. We represent U as a word over the alphabet $\{0,1\}$ of length n having a 1-entry in the j-th position if $j \in U$, i.e., we consider $(\chi^U)^T$. A maximal sequence of consecutive 0s in this word is a **valley** of U. In other words, a valley is an inclusion-wise maximal subset $[l, l+j] \subset U^c$. Accordingly, a maximal sequence of consecutive 1 is called a **hill**. A valley and a hill meet at a **slope**. Thus the number of slopes is the number of occurrences of the patterns 01 and 10 in the word, or in other words, the number of $k \in [n-1]$ with $k \in U$ and $k+1 \notin U$ or vice versa.

We start by looking at $\overline{Q_n}$ for small values of n. For $n = 2$, we have

$$Q_2 = \overline{Q_2} = \mathbb{R}_+ \begin{pmatrix} 0 & 1 \\ 1 & 0 \end{pmatrix}.$$

UNBOUNDED EDGES OF $\overline{Q_3}$ We treat the case $n = 3$ by looking at Figure 3.1. There are two edges starting at each vertex. In fact, with some computation, it can be seen that the unbounded edges containing $M(\iota)$ are

$$M\begin{pmatrix}1\\2\\3\end{pmatrix} + \mathbb{R}_+ M\begin{pmatrix}1\\0\\0\end{pmatrix} = \begin{pmatrix}0&1&2\\1&0&1\\2&1&0\end{pmatrix} + \mathbb{R}_+\begin{pmatrix}0&1&1\\1&0&0\\1&0&0\end{pmatrix} \quad \text{and}$$

$$M\begin{pmatrix}1\\2\\3\end{pmatrix} + \mathbb{R}_+ M\begin{pmatrix}1\\1\\0\end{pmatrix} = \begin{pmatrix}0&1&2\\1&0&1\\2&1&0\end{pmatrix} + \mathbb{R}_+\begin{pmatrix}0&0&1\\0&0&1\\1&1&0\end{pmatrix}; \quad \text{while}$$

$$M\begin{pmatrix}1\\2\\3\end{pmatrix} + \mathbb{R}_+ M\begin{pmatrix}1\\0\\1\end{pmatrix} = \begin{pmatrix}0&1&2\\1&0&1\\2&1&0\end{pmatrix} + \mathbb{R}_+\begin{pmatrix}0&1&0\\1&0&1\\0&1&0\end{pmatrix}$$

is not an edge. This agrees with Proposition 3.3.5, because the sets 100 and 110 are incident to ι while 101 and 010 are not.

Remark 3.3.13. *We need to look at symmetry again. For every $\sigma, \pi \in S(n)$ and $U \subset [n]$ we have:*

(a) *By Remark 3.1.9, the pair $\pi \nearrow U$ defines an edge of $\overline{Q_n}$ if and only if the pair $\pi \circ \sigma \nearrow \sigma^{-1}(U)$ defines an edge of $\overline{Q_n}$.*

(b) *U is incident to π if and only if $\sigma^{-1}(U)$ is incident to $\pi \circ \sigma$.*

(c) *U is over the ridge from a permutation π if and only if $\sigma^{-1}(U)$ is over the ridge from $\pi \circ \sigma$.*

(d) *U^c is over the ridge from a permutation π if and only if $\sigma^{-1}(U)^c$ is over the ridge from $\pi \circ \sigma$.*

The last three are most easily realized by noting that $x \mapsto x \circ \sigma$ is a linear isomorphism of L^n taking $(\Pi^{n-1})^\triangle$ onto itself in such a way that the facet corresponding to a permutation π is mapped to the facet corresponding to $\pi \circ \sigma$, and the vertex corresponding to a set U is mapped to the vertex corresponding to the set $\sigma^{-1}(U)$.

Hence, in the case $n = 3$, we know that the half-line defined by the pair $\pi \nearrow U$ is an edge if and only if π and U are incident. Moreover, the set 101 is over the ridge from ι and 010 is its complement. Actually, it is quite easy to prove in general that if U is over the ridge from π, then the half-line defined by the pair $\pi \nearrow U$ is not an edge of $\overline{Q_n}$.

Proposition 3.3.14. *Let $\pi \in S(n)$ and let $U \subset [n]$ be over the ridge from π. The half-line $M(\pi) + \mathbb{R}_+ M(\chi^U)$ defined by the pair $\pi \nearrow U$ is not an edge of $\overline{Q_n}$.*

Proof. By the remark on symmetry on page 47, it is sufficient to prove the claim for the identical permutation $\iota \in S(n)$. Consider a $k \in [n-1]$ and let $\pi' := \langle k, k+1 \rangle$ be the transposition exchanging k and $k+1$. Further define $U := [k-1] \cup \{k+1\}$. Then $M(\chi^U)$ can be written as a conic combination of vectors defining rays issuing from $M(\iota)$. First consider the ray $M(\pi') - M(\iota)$. With some calculation we see

$$M(\pi') - M(\iota) = \begin{pmatrix} 0 & 1 & -1 & 0 \\ \mathbf{1}^T & 0 & 0 & \mathbf{1}^T \\ -\mathbf{1}^T & 0 & 0 & \mathbf{1}^T \\ 0 & -1 & 1 & 0 \end{pmatrix} \begin{matrix} \\ \Big\} k \\ \\ \Big\} k+1. \end{matrix}$$

For $U_0 = [k]$, we have

3.3. The Convex Set Q_n

$$M(\chi^{U_0}) = \begin{pmatrix} 0 & 1 \\ 1 & 0 \end{pmatrix} \begin{matrix} \} k \\ \} k+1, \end{matrix}$$

and as

$$M(\chi^U) = \begin{pmatrix} 0 & 1 & 0 & 1 \\ 1^T & 1 & 0 & 0^T \\ 0^T & 0 & 1 & 1^T \\ 1 & 0 & 1 & 0 \end{pmatrix} \begin{matrix} \} k \\ \\ \} k+1, \end{matrix} \qquad \text{we obtain}$$

$$M(\chi^U) = M(\chi^{U_0}) + \big(M(\pi') - M(\iota)\big).$$

Hence, no ray over the ridge from $M(\iota)$ is an edge in $P_n + \mathrm{CUT}_n$ containing $M(\iota)$. □

Note: we have also proven that π and U^c do not define an edge of $\overline{Q_n}$ if U^c is over the ridge from π.

UNBOUNDED EDGES OF $\overline{Q_4}$ Because of the symmetry, we consider the edges of $\overline{Q_4}$ containing $M(\iota) = M(\iota^-)$ only. We distinguish the sets U by their number of slopes. Clearly, a set U with a single slope is incident either to ι or to ι^-, and we have already dealt with that case in Remark 3.3.9.

The following sets have two slopes: 0100, 0110, 0010 1011, 1001, and 1101. We only have to consider 1011, 1001, and 1101, because the others are their complements. The first one, 1011, is over the ridge from ι^-, and the last one, 1101, is over the ridge from ι. Therefore, by Lemma 3.3.14, we know that the pairs $\iota \nearrow 1011$ and $\iota \nearrow 1101$ do not define edges of $\overline{Q_4}$. For the remaining set with two slopes, 1001, after some experimenting, one can come up with the matrix

$$C^{1001} := \begin{pmatrix} 0 & 1 & -2 & 1 \\ 1 & 0 & 3 & -2 \\ -2 & 3 & 0 & 1 \\ 1 & -2 & 1 & 0 \end{pmatrix}$$

which satisfies the following properties with C replaced by C^{1001} and U by 1001:

$$C \bullet M(\pi) \geq C \bullet M(\iota) \qquad \text{for all } \pi \neq \iota, \iota^-, \tag{3.5a}$$

$$C \bullet M(\chi^{U'}) \geq 0 \qquad \text{for all } U' \neq U, U^c, \tag{3.5b}$$

$$C \bullet M(\chi^U) < 0. \tag{3.5c}$$

By Farkas' Lemma, the existence of such a matrix C satisfying (3.5) is equivalent to $M(\iota) + \mathbb{R}_+ M(\chi^U)$ being an edge. Another, even simpler, equivalent condition is the existence of a matrix D satisfying the following inequalities:

$$D \bullet M(\pi) > D \bullet M(\iota) \qquad \text{for all } \pi \neq \iota, \iota^-, \tag{3.6a}$$

$$D \bullet M(\chi^{U'}) > D \bullet M(\chi^U) = 0 \qquad \text{for all } U' \neq U, U^c. \tag{3.6b}$$

We find condition (3.5) easier to check for individual matrices, but we will need condition (3.6) in a proof below.

For $n = 4$, we summarize that a pair $\iota \nearrow U$ defines an edge of $\overline{Q_4}$ if and only if U is neither over the ridge from ι nor from ι^-.

UNBOUNDED EDGES OF $\overline{Q_5}$ Let us look at the pairs $\iota \nearrow U$ which define edges in the case $n = 5$. By Remark 3.3.9 and Lemma 3.3.14, we ignore the sets U with one slope and those which are over the ridge from ι or ι^-. When we take only one of each pair of complements, for two slopes, the following list of words remains: 11001, 10011, 10001, 11011. For the last set, consider the matrix C^{11011} in Table A.1 in Appendix A on page 124 satisfying (3.5). It turns out that 11001 can be "reduced to" 1001 by "contracting" the "path" 1–2. Namely, we set

$$C^{11001} := \varepsilon \begin{pmatrix} 0 & 0 & 0 & 0 & 0 \\ 0 & 0 & 1 & -2 & 1 \\ 0 & 1 & 0 & 3 & -2 \\ 0 & -2 & 3 & 0 & 1 \\ 0 & 1 & -2 & 1 & 0 \end{pmatrix} + \begin{pmatrix} 0 & \omega & -1 & -1 & -1 \\ \omega & 0 & 1 & 1 & 1 \\ -1 & 1 & 0 & 0 & 0 \\ -1 & 1 & 0 & 0 & 0 \\ -1 & 1 & 0 & 0 & 0 \end{pmatrix}$$

for a small $\varepsilon > 0$ and a big $\omega \geq 1$. We give the reasoning for the general case in the following lemma. In the same way, 10001 and 10011 can be reduced to 1001 by contracting the paths 2–3 and 4–5 respectively.

Lemma 3.3.15. *Let U_0 be a nonempty proper subset of $[n]$ whose word has the form $a1b$ for two (possibly empty) words a,b. For any $k \geq 0$ define the subset U_k of $[n+k]$ by its word*

$$U_k := a\underbrace{1\ldots 1}_{k+1}b.$$

If the pair $\iota_n \nearrow U_0$ defines an edge of $\overline{Q_n}$, then the pair $\iota_{n+k} \nearrow U_k$ defines an edge of $\overline{Q_{n+k}}$.

Proof. Let $C \in \mathcal{S}\!M(n)$ be a matrix satisfying conditions (3.5) for $U := U_0$. Fix $k \geq 1$ and let $n' := n + k$. We will construct a matrix $C' \in \mathcal{S}\!M(n')$ satisfying (3.5) for $U := U_k$. For a "big" real number $\omega \geq 1$, define a matrix $B_\omega \in \mathcal{S}\!M(k+1)$ whose entries are zero except for those connecting j and $j+1$ for $j \in [k]$:

$$B_\omega := \begin{pmatrix} 0 & \omega & & & 0 \\ \omega & 0 & \omega & & \\ & \omega & & \ddots & \\ & & & & \omega \\ 0 & & & \omega & 0 & \omega \\ & & & & \omega & 0 \end{pmatrix}.$$

We use this matrix to put a heavy weight on the "path" which we "contract." For our second ingredient, let l_a denote the length of the word a and l_b the length of the word b (note that $l_a = 0$ and $l_b = 0$ are possible). Then we define

$$B_- := \begin{pmatrix} -1 & \ldots & -1 \\ 0_{k-1} & \ldots & 0_{k-1} \\ +1 & \ldots & +1 \end{pmatrix} \in \mathbb{R}((k+1) \times l_a) \quad \text{and}$$

$$B_+ := \begin{pmatrix} +1 & \ldots & +1 \\ 0_{k-1} & \ldots & 0_{k-1} \\ -1 & \ldots & -1 \end{pmatrix} \in \mathbb{R}((k+1) \times l_b).$$

Note that 0_{k-1} stands for a column of $k-1$ zeros. Putting these matrices together we obtain a $n' \times n'$-matrix B:

$$B := \begin{pmatrix} 0 & B_-^T & 0 \\ B_- & B_\omega & B_+ \\ 0 & B_+^T & 0 \end{pmatrix}.$$

Now it is easy to check that for any $\pi' \in S(n')$ we have $B \bullet M(\pi') \geq B \bullet M(\iota)$. Moreover, let $\pi' \in S(n')$ satisfy $B \bullet M(\pi') < B \bullet M(\iota) + 1$. By exchanging π' with π'^-, we can assume that $\pi'(1) < \pi'(n')$. It is easy to see that such a π' then has the following "coarse" structure:

$$\pi'([l_a]) \subset [l_a]$$
$$\pi'([n'] \setminus [n' - l_b]) \subset [n'] \setminus [n' - l_b] \quad (3.7)$$
$$\pi'(j) = j \quad \text{for all } j \in \{l_a + 1, \ldots, l_a + k + 1\}.$$

Thus, the matrix B enforces that the "coarse structure" of a $\pi' \in S(n')$ minimizing $B \bullet M(\pi')$ coincides with ι. We now modify the matrix C to take care of the fine structure. For this, we split C into matrices $C_{11} \in \mathcal{S}\!M(l_a)$, $C_{22} \in \mathcal{S}\!M(l_b)$, $C_{12} \in \mathbb{R}(l_a \times l_b)$, $C_{21} = C_{12}^T \in \mathbb{R}(l_b \times l_a)$, and vectors $c \in \mathbb{R}^{l_a}$, $d \in \mathbb{R}^{l_b}$ as follows.

$$C = \begin{pmatrix} C_{11} & c & C_{12} \\ c^T & 0 & d^T \\ C_{21} & d & C_{22} \end{pmatrix}.$$

Then we define the "stretched" matrix $\check{C} \in \mathcal{S}\!M(n')$ by

$$\check{C} := \begin{pmatrix} C_{11} & c & 0 & 0 & C_{12} \\ c^T & 0 & & 0 & 0^T \\ 0 & & 0 & & 0 \\ 0^T & 0 & & 0 & d^T \\ C_{21} & 0 & 0 & d & C_{22} \end{pmatrix}$$

3.3. The Convex Set Q_n

where the middle $\mathbb{0}$ has dimensions $(k-1) \times (k-1)$. Finally, we let $C' := B + \varepsilon \check{C}$ where $\varepsilon > 0$ is small. We show that C' satisfies (3.5).

We first consider $C' \bullet M(\chi^{U'})$ for nonempty subsets $U' \subsetneq [n']$. Note that, if U' contains $\{l_a+1, \ldots, l_a+k+1\}$, then for $U := U' \setminus \{l_a+1, \ldots, l_a+k+1\}$, we have $C' \bullet M(\chi^{U'}) = C \bullet M(\chi^U)$. Thus, we have $C' \bullet M(\chi^{U_k}) = C \bullet M(\chi^{U_0}) < 0$ proving (3.5c) for C' and U_k. For every other U' with $C' \bullet M(\chi^{U'}) < 0$, if ω is big enough, then either U' or U'^c contains $\{l_a+1, \ldots, l_a+k+1\}$, and w.l.o.g. we assume that U' does. By (3.5b) applied to C and U, we know that this implies $U = U_0$ or $U = U_0^c$ and hence $U' = U_k$ or $U'^c = U_k$. Thus, (3.5b) holds for C' and U_k.

Second, we address the property concerning permutations. To show (3.5a), let $\pi' \in S(n)$ be given which minimizes $C' \bullet M(\pi')$. Again, by replacing π' by π'^- if necessary, we assume $\pi'(1) < \pi'(n')$ w.l.o.g. If ε is small enough, we know that π' has the coarse structure displayed in (3.7). This implies that we can define a permutation $\pi \in S(n)$ by letting

$$\pi(j) := \begin{cases} \pi'(j) & \text{if } j \in [l_a] \\ \pi'(j) = j & \text{if } j = l_a + 1 \\ \pi'(j-k) + k & \text{if } j \in [n] \setminus [l_a + 1] \end{cases}.$$

Due to the definition of C' and as $\bullet$ is a linear function, we have

$$C' \bullet M(\pi') - C' \bullet M(\iota_{n'}) = \varepsilon \check{C} \bullet M(\pi') - \varepsilon \check{C} \bullet M(\iota_{n'}) + B \bullet (M(\pi') - M(\iota_{n'})).$$

From the definition of B it follows that

$$B \bullet (M(\pi') - M(\iota_{n'})) \geq 0.$$

As every distance between an element on the first and the last part of the permutation increases by exactly the number k of elements added in the middle of the permutation, we have

$$\varepsilon \check{C} \bullet M(\pi') - \varepsilon \check{C} \bullet M(\iota_{n'}) + B \bullet (M(\pi') - M(\iota_{n'}))$$
$$= \varepsilon \left[C \bullet M(\pi) + k \cdot C \bullet \begin{pmatrix} \mathbb{0}_{l_a \times l_a} & \mathbb{1} \\ \mathbb{1} & \mathbb{0}_{l_b \times l_b} \end{pmatrix} - \left(C \bullet M(\iota_n) + k \cdot C \bullet \begin{pmatrix} \mathbb{0}_{l_a \times l_a} & \mathbb{1} \\ \mathbb{1} & \mathbb{0}_{l_b \times l_b} \end{pmatrix} \right) \right].$$

Altogether, these arguments lead to the following:

$$C' \bullet M(\pi') - C' \bullet M(\iota_{n'}) \geq \varepsilon \left[C \bullet M(\pi) + k \cdot C \bullet \begin{pmatrix} \mathbb{0}_{l_a \times l_a} & \mathbb{1} \\ \mathbb{1} & \mathbb{0}_{l_b \times l_b} \end{pmatrix} \right.$$
$$\left. - \left(C \bullet M(\iota_n) + k \cdot C \bullet \begin{pmatrix} \mathbb{0}_{l_a \times l_a} & \mathbb{1} \\ \mathbb{1} & \mathbb{0}_{l_b \times l_b} \end{pmatrix} \right) \right]$$
$$= \varepsilon \left[C \bullet M(\pi) - C \bullet M(\iota_n) \right]$$
$$\geq 0.$$

Hence, (3.5a) holds. □

Note: the just proven lemma applies to paths of ones too, by exchanging the respective set by its complement.

We come back to $\overline{Q_5}$. The sets with three slopes which are not over the ridge from ι or ι^- are 10110, 10010, and their complements. Lemma 3.3.15 is useless here, since after contraction we would end up with sets which are over the ridge from ι_4 or ι_4^-. However, we can still find matrices satisfying (3.5), they are presented in Table A.1 in Appendix A on page 124. The condition (3.5) can be verified by some case distinctions. The same holds for the unique (up to complement) set with four slopes: 10101. In Table A.1, we offer the matrix C^{10101} satisfying (3.5).

If all valleys and hills of a subset U of $[n]$ consist of only one element (as in 10101) or, equivalently, if U has the maximal possible number $n-1$ of slopes, or, equivalently, if U consists of all odd or all even numbers in $[n]$, we speak of an **alternating set**. Thus (with the trivial exception of the word 10 for $n = 2$), $n = 5$ is the smallest value of n such that for an alternating subset U of $[n]$ the pair $\iota \nearrow U$ defines an edge of $\overline{Q_n}$.

For $n = 5$, we summarize that for all sets U which are not over the ridge from ι or ι^- the pair $\iota \nearrow U$ defines an edge of $\overline{Q_5}$.

Table 3.2: List of all sets with two slopes (up to complement).

	Word			
Hill 1	Valley	Hill 2	Edge?	Why?
1	0	1	no	over the ridge from ι
1	0	1...1	no	over the ridge from ι^-
1	0...0	1	yes	matrix C^{1001} (see Appendix A)
1	0...0	1...1	yes	reduce to 1001
1...1	0	1	no	over the ridge from ι
1...1	0	1...1	yes	matrix C^{11011} (see Appendix A)
1...1	0...0	1	yes	reduce to 1001
1...1	0...0	1...1	yes	reduce to 11011

UNBOUNDED EDGES OF $\overline{Q_6}$ For $n = 6$, we only consider the sets which are not incident to ι, which are not over the ridge from ι or ι^-, which cannot be reduced by Lemma 3.3.15, and which are not complements of any of the already mentioned sets. Only one set remains, namely, the alternating subsets of $\{1,\ldots,6\}$. We give a matrix C^{101010} satisfying (3.5) in Table A.1 in Appendix A. Again we observe that a pair $\iota \nearrow U$ defines an edge of $\overline{Q_6}$ if and only if it is not over the ridge from ι or ι^-.

GENERAL CASE After these preparations for $n \leq 6$, we can consider the general case. The remainder of this section is devoted to a proof of the following theorem.

Theorem 3.3.16. *The edges of $\overline{Q_n}$ containing $M(\pi)$ are precisely the half-lines $M(\pi) + \mathbb{R}_+ M(\chi^U)$ where π is a permutation in $S(n)$ and U is a nonempty proper subset of $[n]$ with the property that neither U nor U^c is over the ridge from π.*

Proof. By Remark 3.1.9, we only need to consider $\pi = \iota$. Again we distinguish the sets U by their number of slopes.

One slope. This is equivalent to U or U^c being incident to ι. We have already treated this case in Proposition 3.3.5, see Remark 3.3.9.

Two slopes. We tackle this case with the preparatory examples above. The complete list of all possibilities, up to complements, and how they are dealt with is summarized in Table 3.2. In this table, 0 stands for a valley consisting of a single zero while $0\ldots0$ stands for a valley consisting of at least two zeros (the same with hills).

Three slopes. This case can also be tackled using the methods we have developed in the examples. Table 3.3 presents the results.

An even number $s \geq 4$ of slopes. Using Lemma 3.3.15, we reduce such a set to an alternating set with s slopes showing that for all these sets U the pair $\iota \nearrow U$ defines an edge of $\overline{Q_n}$. This is in accordance with the statement of the theorem because sets which are over the ridge from ι can have at most three slopes. They can therefore not occur in any other case. The statement for alternating sets is proven by induction on n in Lemma 3.3.17 below. Note that the start of the induction is $n = 5$, which we dealt with in the examples above.

An odd number $s \geq 5$ of slopes. Again, using Lemma 3.3.15, we reduce such a set to an alternating set with s slopes and refer to Lemma 3.3.18 to perform the induction starting with example $n = 6$. □

Lemma 3.3.17. *For an odd integer $n \geq 5$, let U be an alternating subset of $[n]$. The pair $\iota \nearrow U$ defines an edge of $\overline{Q_n}$.*

Proof. The proof is by induction over n. From the example for $n = 5$ above, the start of the induction is guaranteed. Moreover, there exists a $D^5 \in S_sM(5)$ satisfying (3.6). We will need this matrix in the inductive construction.

Table 3.3: List of all sets with three slopes (up to complement).

Word				Edge?	Why?
Hill 1	Valley 1	Hill 2	Valley 2		
1	0	1	0	no	over the ridge from ι
1	0	1	0...0	no	over the ridge from ι
1	0	1...1	0	yes	matrix C^{10110} (see Appendix A)
1	0	1...1	0...0	yes	reduce to 10110
1	0...0	1	0	yes	matrix C^{10010} (see Appendix A)
1	0...0	1	0...0	yes	reduce to 10010
1	0...0	1...1	0	yes	reduce to 10010
1	0...0	1...1	0...0	yes	reduce to 10110
1...1	0	1	0	no	over the ridge from ι
1...1	0	1	0...0	no	over the ridge from ι
1...1	0	1...1	0	yes	reduce to 10110
1...1	0	1...1	0...0	yes	reduce to 10110
1...1	0...0	1	0	yes	reduce to 10010
1...1	0...0	1	0...0	yes	reduce to 10010
1...1	0...0	1...1	0	yes	reduce to 10010
1...1	0...0	1...1	0...0	yes	reduce to 10010

Now assume that the pair ι, U^- defines an edge of $\overline{Q_n}$ where U^- is an alternating subset of $[n]$. W.l.o.g., we assume that $U^- = 10\ldots01$. There exists a matrix $D^- \in \mathbb{S}M(n)$ for which (3.6) holds. We will construct a matrix $D \in \mathbb{S}M(n+2)$ satisfying (3.6) for $U := 010\ldots010$.

We extend D^- to a $(n+2) \times (n+2)$-Matrix

$$\widehat{D} := \begin{pmatrix} D^- & 0 & 0 \\ 0^T & 0 & 0 \\ 0^T & 0 & 0 \end{pmatrix}.$$

We do the same with D^s, only on the other side:

$$\widehat{D^s} := \begin{pmatrix} 0 & 0 & 0^T \\ 0 & 0 & 0^T \\ 0 & 0 & D^s \end{pmatrix}$$

Now we let $D := \widehat{D} + \widehat{D^s}$ and check the conditions (3.6) on D, which can now easily be verified. □

The even case is prooven in the same way.

Lemma 3.3.18. *For an even integer $n \geq 6$, let U be an alternating subset of $[n]$. The pair $\iota \nearrow U$ defines an edge of $\overline{Q_n}$.*

Our theorem states that the unbounded edges of $\overline{Q_n}$ starting at $M(\pi)$ are precisely the rays issuing from $M(\pi)$ in the direction $M(\chi^U)$ for sets U, $\emptyset \neq U \subsetneq [n]$, which are *not* over the ridge from π. From Theorem 3.3.16, we immediately have the following results.

Corollary 3.3.19. *For $n \geq 4$, the number of unbounded edges issuing from a vertex of $\overline{Q_n} = P_n + \text{CUT}_n$ is $2^{n-1} - n$.*

Proof. The number of nonempty proper subsets of $[n]$ is $2^n - 2$. There are precisely $n-1$ sets which are over the ridge from π and the same number for π^-. Since the mapping M identifies antipodal points, we obtain the number given above. □

Corollary 3.3.20. $\overline{Q_n} = Q_n \Leftrightarrow n \leq 3$.

Proof. Consider $n = 2$. As there are only two permutations ι and $\bar{\iota}$ whose corresponding set $U_\emptyset$ is trivially incident to ι, Theorem 3.3.16 says that there is exactly one extreme ray $\mathbb{R}_+ M(\chi^{U_0})$ containing $M(\iota)$. This corresponds to Proposition 3.3.5, therefore $Q_2 = P_2 + C_2 = \overline{Q_2}$. For $n = 3$, consider $\pi_1 := (1\ 2\ 3)$, $\pi_2 := (1\ 3\ 2)$, and $\pi_3 := (2\ 1\ 3)$. All other permutations are the antipodal permutations of these. The sets corresponding to π_1 are $U_1 = \{1\}$ and $U_2 = \{1,2\}$. As both are incident to ι, we know, by Theorem 3.3.16, that $M(\chi^{U_1})$ and $M(\chi^{U_2})$ are extreme rays of $P_3 + C_3$ containing $M(\iota)$. The sets corresponding to π_2 are U_1 and $U_3 = \{1,3\}$, for π_3 we have U_3 and U_2. As U_3 is not incident to ι, it does not induce an extreme ray for $P_3 + C_3$ containing $M(\iota)$. Again, this is in accordance with Proposition 3.3.5, therefore we have $Q_3 = P_3 + C_3 = \overline{Q_3}$.

For all $n \geq 4$, Proposition 3.3.5 and Theorem 3.3.16 are in conflict, hence $Q_n \neq P_n + \text{CUT}_n = \overline{Q_n}$. □

3.3.4 Inequalities Defining Facets of $\overline{Q_n}$

In the following, we study linear inequalities that define facets of $\overline{Q_n}$, i.e., faces of dimension $\binom{n}{2} - 1$. We first present some general results about such inequalities and then list some specific inequalities.

All results presented in this section were obtained by A. Letchford. They have been published in our paper [102] and are mentioned here for the sake of completeness.

General Results on Facet-Defining Inequalities

In this subsection, we prove a structural result about inequalities that define facets of $\overline{Q_n}$, and show how this can be used to construct facets of $\overline{Q_n}$ in a mechanical way from facets of either P_n or CUT_n.

3.3. The Convex Set Q_n

We will need the following definition taken from [45]. Let $a^T x \geq b$ be a linear inequality where $a, x \in \mathbb{R}^{\binom{n}{2}}$. The inequality is said to be **canonical** if

$$\min_{\emptyset \neq S \subset [n]} \sum_{i \in S} \sum_{[n] \setminus S} a_{ij} = 0. \tag{3.8}$$

By definition, an inequality $a^T x \geq 0$ defines a proper face of CUT_n if and only if it is canonical. In [104], it is shown that every facet of P_n is defined by a canonical inequality. The following lemma is the analogous result for $\overline{Q_n}$.

Lemma 3.3.21. *Every unbounded facet of $\overline{Q_n}$ is defined by a canonical inequality.*

Proof. Suppose that the inequality $a^T x \geq b$ defines an unbounded facet of $\overline{Q_n}$. Since $\overline{Q_n}$ is the Minkowski sum of P_n and CUT_n, the inequality must be valid for CUT_n. Therefore, the left-hand side of (3.8) must be non-negative. Moreover, since the inequality defines an unbounded facet, there must be at least one extreme ray of CUT_n satisfying $b^T x = 0$. Therefore, the left-hand side of (3.8) cannot be positive. □

We remind the reader that only one facet of $\overline{Q_n}$ is bounded, which is P_n.

Now, we show how to derive facets of $\overline{Q_n}$ from facets of P_n.

Proposition 3.3.22. *Let F be any facet of P_n, and let $a^T x \geq b$ be the canonical inequality that defines it. This inequality defines a facet of $\overline{Q_n}$ as well.*

Proof. The fact that the inequality is valid for $\overline{Q_n}$ follows from the fact that $\overline{Q_n}$ is the Minkowski sum of P_n and CUT_n. Now, since F is a facet of P_n, there exist $\binom{n}{2} - 1$ affinely-independent vertices of P_n that satisfy the inequality with equality. Moreover, since the inequality is canonical, there exists at least one extreme ray of CUT_n that satisfies $a^T x = 0$. Since $\overline{Q_n}$ is the Minkowski sum of P_n and CUT_n, there exist $\binom{n}{2}$ affinely independent points in $\overline{Q_n}$ that satisfy the inequality with equality. Thus, the inequality defines a facet of $\overline{Q_n}$. □

Now, we show how to derive facets of $\overline{Q_n}$ from facets of CUT_n.

Proposition 3.3.23. *Let $a^T x \geq 0$ define a facet of CUT_n, and let b be the minimum of $a^T x$ over all $x \in P_n$. Then the inequality $a^T x \geq b$ defines a facet of $\overline{Q_n}$.*

Proof. As before, the fact that the inequality $a^T x \geq b$ is valid for $\overline{Q_n}$ follows from the fact that $\overline{Q_n}$ is the Minkowski sum of P_n and CUT_n. Now, since the inequality $a^T d \geq 0$ defines a facet of CUT_n, there exist $\binom{n}{2} - 1$ linearly independent extreme rays of CUT_n that satisfy $a^T x = 0$. Moreover, from the definition of b, there exists at least one extreme point of P_n that satisfies $b^T x = b$. Since $\overline{Q_n}$ is the Minkowski sum of P_n and CUT_n, there exist $\binom{n}{2}$ affinely-independent points in $\overline{Q_n}$ that satisfy $a^T x = b$. Thus, the inequality $a^T x \geq b$ defines a facet of $\overline{Q_n}$. □

Some Specific Facet-Defining Inequalities

The results in the previous section enable one to derive a wide variety of facets of $\overline{Q_n}$. In this section, we briefly examine some specific valid inequalities; namely, the inequalities mentioned in [45]. First, we deal with the clique and pure hypermetric inequalities.

Proposition 3.3.24. *The clique inequalities, see Section 1.5.2, define facets of $\overline{Q_n}$ for all $S \subseteq [n]$ with $|S| \geq 2$.*

Proof. It was shown in [45] that the clique inequalities define facets of P_n when S is a proper subset of $[n]$. In this case, the inequalities are canonical and so, by Proposition 3.3.22, they define facets of $\overline{Q_n}$ as well. The case $S = [n]$ is covered by the fact that P_n is the only bounded facet of $\overline{Q_n}$. □

Proposition 3.3.25. *All pure hypermetric inequalities define facets of $\overline{Q_n}$.*

Proof. It was shown in [14] that all pure hypermetric inequalities define facets of CUT_n. It was also shown in [45] that every pure hypermetric inequality is satisfied with equality by at least one extreme point of P_n. The result then follows from Proposition 3.3.23. □

As for the strengthened pure negative-type and strengthened star inequalities, it was shown in [45] that they define facets of P_n under certain conditions. Since they are canonical, they define facets of $\overline{Q_n}$ under the same conditions. In fact, using the same proof technique used in [45], one can show the following two results.

Proposition 3.3.26. *All strengthened pure negative-type inequalities define facets of $\overline{Q_n}$.*

Proposition 3.3.27. *Strengthened star inequalities define facets of $\overline{Q_n}$ if and only if $|S| \neq 4$.*

We omit the proofs, for the sake of brevity.

3.4 0/1 Polytope D_n

In this section we look at some polyhedral aspects of the 0/1 problem formulation introduced in this thesis. We start with the definition and basic properties such as the dimension of the polytope. We then present a characterization of some facets of small polytopes.

3.4.1 Definition of D_n and Basic Properties

A vector $d \in \{0,1\}^{\binom{n}{2}(n-1)}$ and a permutation $\pi \in S(n)$ **correspond** to each other if

$$d_{ijk} = \begin{cases} 1, & |\pi(i) - \pi(j)| = k, \\ 0, & \text{otherwise} \end{cases} \quad \text{for all } i < j \in V \text{ and } k = 1, \ldots, n-1.$$

We denote the vector corresponding to π with d^π. If no confusion can arise, we omit the exponent. We will now define the polytope D_n corresponding to the MinLA model based on d-variables:

$$D_n := \text{conv}\{d \in \{0,1\}^{\binom{n}{2}(n-1)} \mid \text{exists } \pi \in S(n) \text{ such that } d = d^\pi\}. \tag{3.9}$$

Obviously, D_n is bounded and its vertices are the vectors corresponding to permutations, i.e., $\text{vert}(D_n) = \{d \mid \text{exists } \pi \in S(n) \text{ such that } d = d^\pi\}$. The polytope P_n is a projection of D_n:

$$f : D_n \longrightarrow P_n$$
$$d^\pi \mapsto y^\pi.$$

It is therefore clear that all valid inequalities of P_n are valid for D_n. The big question is which facets of P_n are facets of D_n as well. To investigate this, we first need to state the dimension of D_n. Unfortunately, D_n is not full-dimensional and has a very complex structure. Nevertheless we will present a conjectured minimal equation system of this polytope.

3.4.2 Conjectured Minimal Equation System

Conjecture 3.4.1. *The minimal equation system of D_n is*

$$\sum_{k=1}^{n-1} d_{ijk} = 1 \quad \text{for all } i < j \in V, \tag{3.10}$$

$$\sum_{i<j} d_{ijk} = n - k \quad \text{for all } k = 1, \ldots, n-2, \tag{3.11}$$

$$\sum_{i \neq j} d_{ijk} + d_{ij(n-k)} = 2 \quad \text{for all } j < n, k < \lfloor n/2 \rfloor, n > 3. \tag{3.12}$$

This leads directly to the next conjecture about the dimension of the polytope.

3.4. 0/1 Polytope D_n

Conjecture 3.4.2. *The conjectured dimension of the polytope is*

$$\binom{n}{2}(n-1) - \Big((n-2) + (n-1)\big(n/2 + \lfloor n/2 \rfloor - 1\big)\Big). \tag{3.13}$$

We will show that constraints (3.10), (3.11), and (3.12) are linearly independent. In the following, we consider a constraint in its general form $\sum_{i<j\in V}\sum_{k=1}^{n-1} a_{ijk} d_{ijk}$. Furthermore, we define a **block** corresponding to a pair of nodes $i, j \in V$ to be the coefficients a_{ijk} for all $k = 1,\ldots,n-1$.

Lemma 3.4.3.

(a) All constraints of type (3.10) and (3.11) are linearly independent.

(b) All constraints of type (3.10), (3.11), and (3.12) are linearly independent.

Proof. (a). Assume there exists a constraint

$$\sum_{k=1}^{n-1} d_{ijk} = 1 \tag{3.14}$$

of type (3.10) which is a linear combination of (3.11) and other (3.10) constraints. Its coefficients a_{ijk} are equal to 1 for all k and 0 otherwise. The coefficient $a_{ij(n-1)}$ equals 0 in all other (3.10) constraints. Furthermore, $a_{ij(n-1)}$ is not part of any (3.11) constraint, hence no linear combination of (3.14) is possible.

Now assume there exists a constraint

$$\sum_{i<j} d_{ijl} = n - l \tag{3.15}$$

of type (3.11) which is a linear combination of (3.10) and other (3.11) constraints. Its coefficients a_{ijl} are equal to 1 for all pairs $i < j \in V$ and 0 otherwise. As non-zero entries at these positions can only come from type (3.10) constraints, we know that the linear combination (3.15) must contain the sum of all (3.10) constraints. Considering this sum, we have far too many 1-entries. Some of them can be deleted by subtracting constraints of type (3.11) for $k \neq l, n-1$. But the coefficients $a_{ij(n-1)}$ cannot be deleted, and as $l \neq n-1$, constraint (3.15) cannot be a linear combination of (3.11), and (3.10).

(b). Assume there exists a constraint of type (3.10) which is a linear combination of (3.11), (3.12) and other (3.10) constraints. Let

$$\sum_{k=1}^{n-1} d_{ijk} = 1 \tag{3.16}$$

be this constraint. All its coefficients a_{ijl} are 0 except the ones in the block corresponding to $i < j \in V$. Considering the other constraints, we see that the non-zero coefficients $a_{ij(n/2)}$ (for n even) and $a_{ij(n/2\pm1)}$ (for n odd) can only come from type (3.11) constraints. But as these constraints have non-zero coefficients in *all* blocks, constraint (3.16) cannot be a linear combination.

Now assume there exists a constraint

$$\sum_{i<j} d_{ijl} = n - l \tag{3.17}$$

of type (3.11) which is a linear combination of (3.10), (3.12), and other (3.11) constraints. Due to (a), it is clear that type (3.12) constraints have to be part of the linear combination of (3.17). Considering (3.12) more closely, we note several properties. First, a block corresponding to $i < j < n$ has non-zero coefficients in exactly two constraints. Second, all blocks corresponding to $i < j = n$ occur only once with non-zero coefficients. Third, it is impossible to sum up all constraints such that each block occurs exactly once. For the linear combination, coefficients in all blocks have to be changed. Therefore, the constraints of type (3.12) cannot be part of the linear combination of (3.17).

At last assume there exists a constraint of type (3.12) which is a linear combination of (3.10), (3.11), and other (3.12) constraints. Let

$$\sum_{i \neq j} d_{ijl} + d_{ij(n-l)} = 2 \tag{3.18}$$

be this linearly dependent constraint. Note that the structure of each block can be constructed by summing up all (3.10) constraints and subtracting certain (3.11) constraints. In this way, the structure is constructed simultaneously for all $n(n-1)/2$ blocks. Note that this structure cannot be obtained for less blocks using constraints of type (3.10) and (3.11). As in (3.18) exactly $n-1$ blocks have non-zero coefficients, we have to delete several of the $n(n-1)/2$ non-zero blocks. This can only be done with constraints of type (3.12). But from the definition of these constraints it follows directly that it is impossible to delete the non-zero coefficients of only these blocks. □

From Lemma 3.4.3, we have the following consequence.

Corollary 3.4.4. *Expression (3.13) is an upper bound for the dimension of D_n.*

3.4.3 SMALL POLYTOPES

With the help of the POlyhedron Representation Transformation Algorithm PORTA [108], the complete linear description of all polytopes D_n for $n \leq 5$ can be computed explicitly. Due to the high complexity of the structure of the polytope it was not possible to compute the outer description of D_n for larger instances. This gives us only few hints for the search of facet defining inequalities. The situation is even more complicated as we have at least three different types of equations in the minimal equation system. Therefore one single facet can be presented in very different ways and is therefore hard to recognize.

We will now present some facets of the polytopes of very low dimension. For $n = 3$, the polytope is described by three lower-bounding inequalities. For $n = 4$, eight lower-bounding inequalities are facets. We will explain the remaining four. The first inequality

$$d_{2\,4\,2} + d_{3\,4\,2} \leq 1$$

is the upper-bounding inequality $d_{1\,4\,2}$, as it can be substracted from the equation

$$d_{1\,4\,2} + d_{2\,4\,2} + d_{3\,4\,2} = 1.$$

The second inequality

$$d_{3\,4\,2} \leq d_{1\,3\,3} + d_{1\,4\,3} + d_{2\,3\,3} + d_{2\,4\,3}$$

describes the following situation. If nodes 3 and 4 have a distance two, then at least one of the node pairs 1,3; 1,4; 2,3 or 2,4 has to have a distance three. To understand the inequality

$$d_{1\,4\,3} + d_{2\,3\,3} \leq d_{2\,4\,2} + d_{3\,4\,2}$$

we consider two cases. If nodes 1 and 4 are the left- and rightmost nodes, the nodes 2 and 3 have to be the middle nodes. In the other case, nodes 2 and 3 are the left- and rightmost nodes, therefore node 4 has to be one of the middle nodes. For the fourth facet-defining inequality

$$d_{1\,3\,3} + d_{2\,4\,2} + d_{2\,4\,3} \leq 1$$

we use the each-edge-one-distance equality $d_{2\,4\,1} + d_{2\,4\,2} + d_{2\,4\,3} = 1$ to simplify it. We obtain

$$d_{1\,3\,3} \leq d_{2\,4\,1}.$$

which requires nodes 2 and 4 to be the middle nodes if nodes 1 and 3 are the left- and rightmost nodes. A generalization of this inequality would e. g. be the following

$$d_{i'j'(n-1)} + \sum_{i,j \in V \setminus \{i',j'\}} \left(d_{ij(n-1)} + d_{ij(n-2)} \right) \leq 1, \qquad \text{for all } i',j' \in V. \tag{3.19}$$

Here, we choose one pair of nodes i' and j' and consider them as the left- and rightmost nodes. For all other pairs of nodes the distances $n-1$ and $n-2$ are not possible anymore. We tried to find this inequality within the 8045 facet-defining inequalities of D_5 but could not identify them. Note: this does not mean that (3.19) is not facet-defining, as it might be presented in a "hidden" form due to the three different equation types. Most of the facet-defining inequalities of D_5 have a complex structure. Nevertheless, we could identify 14 inequalities as lower-bounding inequalities.

3.5 Integral Distance and Assignment Polyhedron P_n^A

In this section we focus on the polyhedron associated with the revisited assignment formulation of the MinLA. We start with some basic properties and trivial facets. Furthermore, we present how some inequalities investigated in [45] can be strengthened by *lifting in* the x-variables.

3.5.1 Definition of P_n^A and Basic Properties

The improved assignment formulation, see Section 2.3 on page 42, allows y_{ij} to be larger than $|\pi(i) - \pi(j)|$. Potentially, one could associate an unbounded polyhedron with it. However, we prefer to work with a polytope and therefore would like y_{ij} to be *equal to* $|\pi(i) - \pi(j)|$. Equivalently, we will require the following non-linear constraints to hold

$$y_{ij} = \left| \sum_{p=1}^{n} p(x_{ip} - x_{jp}) \right| \qquad \text{for all } i,j \in V. \tag{3.20}$$

Accordingly, we define the following polytope

$$P_n^A = \operatorname{conv}\left\{ (x,y) \in \{0,1\}^{n^2} \times \mathbb{Z}_+^{\binom{n}{2}} \mid (2.49), (2.50), (3.20) \text{ hold} \right\}. \tag{3.21}$$

We will see later that it is possible to define P_n^A using linear inequalities only (rather than using the non-linear constraint (3.20)).

Dimension and Trivial Facets

Clearly, P_n^A is not full-dimensional due to the presence of the assignment Constraints (2.49) and (2.50). (Only $2n-1$ of the assignment constraints are linearly independent.) Moreover, we have n further equations as follows:

$$\sum_{j \in V, j \neq i} y_{ij} = \left\lfloor \frac{n^2}{4} \right\rfloor + \sum_{p=1}^{\lfloor (n-1)/2 \rfloor} \left\lfloor \left(\frac{n+1}{2} + p\right)^2 \right\rfloor (x_{ip} + x_{i(n-p+1)}) \qquad i \in V. \tag{3.22}$$

Note that these equations, together with the constraints (2.50), imply the equation $\sum_{i<j\in V} y_{ij} = \binom{n+1}{3}$. We conjecture that these $3n-1$ equations are independent and completely describe the affine hull. We conjecture further, that the non-negativity inequalities $x_{ip} \geq 0$ for all i and p induce facets. The trivial bounds $1 \leq y_{ij} \leq n-1$, on the other hand, do not induce facets (see the following sections).

3.5.2 Bounding Inequalities

We have found some inequalities that impose lower and upper bounds on the y-variables in terms of the x-variables. We present the lower-bounding inequalities first, since they are likely to be more useful as cutting planes than the upper-bounding inequalities (given the nature of the objective function).

Lower-Bounding Inequalities

The lower-bounding inequalities can be viewed as a generalization of the constraints (2.54) and (2.55). They are presented in the following lemma:

Lemma 3.5.1. *The lower-bounding inequalities*

$$y_{ij} \geq \sum_{q=1}^{n} |p-q|(x_{iq} - x_{jq}) \quad \text{for all } i < j \in V, p = 1, \ldots, n \tag{3.23}$$

are valid for P_n^A.

Proof. The inequality is satisfied with equality when either $\pi(i) < \pi(j) \leq p$ or $\pi(i) > \pi(j) \geq p$ and has a positive slack otherwise. □

When $p \in \{1, n\}$, the lower-bounding inequalities are equivalent (via the assignment constraints) to the constraints (2.54) and (2.55).

Using the software PORTA, we have found that the lower-bounding inequalities induce faces of high dimension but not quite facets. When $p = 1$, every vector lying on the face also satisfies the equations $x_{i1} = x_{jn} = 0$. Similarly, when $p = n$, they satisfy the equations $x_{j1} = x_{in} = 0$. Finally, when $1 < p < n$, they satisfy the equations $x_{ip} = x_{j1} = x_{jn} = 0$. We believe that one can obtain facets by lifting in the involved variables.

UPPER-BOUNDING INEQUALITIES

The upper-bounding inequalities are similar.

Lemma 3.5.2. *The upper-bounding inequalities*

$$y_{ij} \leq \sum_{q=1}^{n} |p-q|(x_{iq} + x_{jq}) \quad \text{for all } i < j \in V, p = 1, \ldots, n \tag{3.24}$$

are valid for P_n^A.

Proof. The inequality is satisfied with equality when either $\pi(i) \leq p \leq \pi(j)$ or $\pi(i) \geq p \geq \pi(j)$ and has a positive slack otherwise. □

The upper-bounding inequalities for $p = 1$ and $p = n$ do not, however, induce facets of P_n^A. For example, the upper-bounding inequality for $p = 1$ is dominated by the upper-bounding inequality for $p = 2$, the assignment constraint $\sum_{k=1}^{n} x_{k1} = 1$, the assignment constraints $\sum_{q=1}^{n} x_{ij} = 1$ and $\sum_{q=1}^{n} x_{ij} = 1$, and the non-negativity inequalities $x_{k1} \geq 0$ for $k \in V \setminus \{i, j\}$. Thus, it is redundant.

However, we make the following conjecture.

Conjecture 3.5.3. *The upper-bounding inequalities induce facets of* P_n^A *when* $2 \leq p \leq n-1$.

Remark 3.5.4. *We think that the upper-bounding constraints (3.24), together with Constraints (2.49), (1.12), (2.54), and (2.55), imply the non-linear Constraints (3.20). Therefore, they give a full mixed integer programming formulation of the MinLA.*

FURTHER IDEAS

The upper- and lower-bounding inequalities involve only one d-variable. There are other interesting inequalities of this type. For example, the trivial constraints $y_{ij} \geq 1$ are dominated by

$$y_{ij} + \sum_{p \text{ odd}} x_{ip} - \sum_{p \text{ even}} x_{jp} \geq 1$$

and

$$y_{ij} - \sum_{p \text{ odd}} x_{ip} + \sum_{p \text{ even}} x_{jp} \geq 1.$$

A generalization of these inequalities is presented in Section 3.5.3.

One could perhaps find more inequalities of this type by using PORTA to compute the projection of P_n^A onto the subspace defined by the variables that involve two particular vertices i and j. This projection seems to be related to the classical 'cyclic group' polyhedra of Gomory.

3.5.3 Strengthening the Inequalities from Amaral and Letchford 2008

In [45], four classes of facet-inducing inequalities were presented: the clique, pure hypermetric, bipartite and strong star inequalities. In this section, we show that all these inequalities can be strengthened by *lifting in* the x-variables.

Strengthened Hypermetric Inequalities

Recall the definition of a cut vector and the cut cone described in Section 0.3.3. The following result is a simplified version of the result in [45]:

Proposition 3.5.5. *Let $\pi \in S(n)$ be a permutation and let $y^\pi \in \mathbb{R}_+^{\binom{n}{2}}$ be the corresponding distance vector. Then y^π is the sum of $n-1$ cut vectors. Specifically,*

$$y^\pi = \sum_{k=1}^{n-1} y^k,$$

where, for $k = 1,\ldots,n-1$, the cut vector y^k is formed by setting U to the set of vertices placed in the first k positions (or, more formally, to $\{\pi^{-1}(1),\ldots,\pi^{-1}(k)\}$).

Thus, all valid inequalities for the cut cone are valid for P_n^A. As in [45], we consider the pure hypermetric inequalities, which induce facets of the cut cone. These take the form

$$\sum_{i,j \in V} b_i b_j y_{ij} \leq 0 \quad \text{where} \sum_{i \in V} b_i = 1 \text{ and } b_i \in \{0, \pm 1\} \text{ for all } i \in V.$$

They can also be written in the alternative form

$$\sum_{i \in S, j \in T} y_{ij} - \sum_{ij \in E(T)} y_{ij} - \sum_{ij \in E(S)} y_{ij} \geq 0 \quad \text{for all } S, T \subset V, S \cap T = \varnothing, |T| = |S| - 1,$$

where $S = \{i \in V \mid b_i = 1\}$ and $T = \{i \in V \mid b_i = -1\}$.

It follows from results in [45] that, when only y-variables are present, the pure hypermetric inequalities induce facets if and only if $n \geq |S| + |T| + 2$. When the x-variables are present, however, this is no longer true. Indeed, we have the following proposition.

Proposition 3.5.6. *The pure hypermetric inequalities can be strengthened to*

$$\sum_{i,j \in V} b_i b_j y_{ij} \leq -2 \sum_{i \in T} (x_{i1} + x_{in}).$$

Proof. By symmetry, it suffices to show validity for the identity arrangement. Let y^* be the associated distance vector. From the above proposition, we have

$$y^* = \sum_{k=1}^{n-1} y^k,$$

where y^k for $k = 1,\ldots,n-1$ is the cut vector obtained by setting $U = \{1,\ldots,k\}$. Now, we have

$$\sum_{i,j \in V} b_i b_j y_{ij}^* = \sum_{k=1}^{n-1} \sum_{i,j \in V} b_i b_j y_{ij}^k$$

$$= \sum_{k=1}^{n-1} \left(\sum_{i=1}^{k} b_i \sum_{j=k+1}^{n} b_j \right)$$

$$= \sum_{k=1}^{n-1} \left(\sum_{i=1}^{k} b_i \right) \left(1 - \sum_{i=1}^{k} b_i \right).$$

Each of the $n-1$ terms in the outer summation is non-positive. Moreover, if a vertex in T is in position 1, the first of those terms becomes $-1 \times 2 = -2$. Similarly, if a vertex in T is placed in position n, the last of those terms becomes $2 \times -1 = -2$. □

Our experiments with PORTA suggest the following conjecture.

Conjecture 3.5.7. *The strengthened pure hypermetric inequalities induce facets if and only if $n \geq |S| + |T| + 3$.*

Strengthened Bipartite Inequalities

The following bipartite inequalities are similar to the hypermetric inequalities described above.

$$\sum_{i \in S, j \in T} y_{ij} - \sum_{ij \in E(S)} y_{ij} - \sum_{ij \in E(T)} y_{ij} \geq |S| \quad \text{for all } S, T \subset V,$$

$$S \cap T = \emptyset, |T| = |S|,$$

where as before $S = \{i \in V \mid b_i = 1\}$ and $T = \{i \in V \mid b_i = -1\}$. Note that the bipartite inequalities reduce to trivial lower bounds $y_{ij} \geq 1$ when $|S| = |T| = 1$.

It can be shown that the bipartite inequalities induce facets when one works in the space of the y-variables. When the x-variables are present, however, they can be strengthened.

Conjecture 3.5.8. *Let $O = \{p \in \{1,\ldots,n\} \mid p \text{ odd}\}$ and $\mathcal{E} = \{p \in \{1,\ldots,n\} \mid p \text{ even}\}$. For any disjoint S and T with $|S| = |T|$, the following strengthened bipartite inequalities are valid for P_n^A.*

$$\sum_{i,j \in V} b_i b_j y_{ij} \leq -|S| + \sum_{i \in S} \sum_{p \in O} x_{ip} - \sum_{i \in T} \sum_{p \in \mathcal{E}} x_{ip}$$

$$\sum_{i,j \in V} b_i b_j y_{ij} \leq -|S| - \sum_{i \in S} \sum_{p \in O} x_{ip} + \sum_{i \in T} \sum_{p \in \mathcal{E}} x_{ip}$$

Observe that, when n is odd, the second strengthened bipartite inequality looks somehow *stronger* than the first, in the sense that it has more negative coefficients on the left hand side and fewer positive coefficients. Indeed, when $n = 5$ and $|S| = 1$, the first inequality has only 60 roots, whereas the second one has 72 roots.

Our experiments with PORTA also suggest the following conjecture.

Conjecture 3.5.9. *To avoid degenerate cases assume that $n \geq 4$. The strengthened bipartite inequalities (of either kind) induce facets if and only if $n \geq |S| + |T| + 2$.*

Strengthened Star Inequalities

In other papers, we considered the star inequalities

$$\sum_{rj \in E} y_{rj} \geq \left\lfloor \frac{|S|^2}{4} \right\rfloor \quad \text{for every star } (S, E),$$

where r is the center node of the star. They induce facets of the dominant in the y space but not for the polytope in y space. Moreover, when the x-variables are present, they can be easily strengthened by adding the following quantity to the right-hand side

$$\sum_{p=1}^{\lfloor |S|/2 \rfloor} \left\lfloor \frac{|S|}{2} + 1 - p \right\rfloor^2 \left(x_{ip} + x_{i(n-p+1)} \right).$$

When $\lfloor n/2 \rfloor < |S| < n - 1$, the star inequalities can be strengthened in a different way. For example, if $n = 6$, $i = 1$, and $S = \{2,3,4,5\}$, we have

$$\sum_{j \in S} y_{ij} \geq 6 + (x_{62} + x_{65}) + 2(x_{63} + x_{64}).$$

It would be nice to find one or more classes of facets that dominate the star inequalities.

Chapter 4

Branch-and-Cut-and-Price Algorithm

In this chapter we describe the branch-and-cut-and-price algorithm that we have implemented for the MinLA. The key idea is to combine pricing with a traditional branch-and-cut algorithm. The formulation of the MinLA described in Section 2.1 on page 31 is particularly useful for such an algorithm: The first reason for this is that it has a sparse solution structure, i.e., only $\binom{n}{2}$ of $(n-1)\binom{n}{2}$ variables are non-zero in a feasible solution. The second reason is that due to the definition of the d-variables, we can make use of several logical implications during the branching process. In the constraint system all coefficients are known, hence no lifting is necessary.

A flowchart that presents the procedure of a branch-and-cut-and-price algorithm is shown in Figure 2 on page 14. We now describe the way we realized this algorithm for the MinLA.

4.1 Choice of Start Variables

We start the branch-and-cut-and-price algorithm with a small set of variables. These variables correspond to a feasible solution $\pi \in S(n)$ in the following way. We generate all variables d_{ijk} for $|\pi(i) - \pi(j)| = k$. To have more tolerance we additionally generate the variables $d_{ijk'}$ with $|\pi(i) - \pi(j)| = k'$ for $|k' - k| \leq 4$. We can let π be a random or the identity permutation. As it is of great importance that the right variables are in the system from the beginning, we can choose π corresponding to the best solution found by one of the following start heuristics.

4.2 Start Heuristics

4.2.1 Simulated Annealing

We use a standard simulated annealing (**SA**) approach to obtain a feasible arrangement of the nodes, see [34]. It starts with a randomly chosen start solution and generates a series of other solutions. In the beginning worse solutions are accepted which reduces the probability to be caught in a local minimum.

4.2.2 Multi-Start Local Search Routine

The following heuristic is a combination of different procedures and was developed by G. Reinelt [109]. We call it multi-start local search routine (**MLS**). It starts with a random permutation. As long as the current objective function value is not better than the start value we call the following heuristics in the presented ordering.

- Node Insertion
- Subsequence Reversal

Figure 4.1: Example for 2-exchange.

- Local Enumeration
- 2-Exchange
- Kernighan-Lin based on node insertion moves
- Kernighan-Lin based on 2-Exchange moves

2-Exchange

The idea is to change the positions of two nodes if many edge lengths can be shortened by the exchange step. Hence, we consider all nodes adjacent to u and v. If most of the nodes adjacent to u lie on the other side of v and most of the nodes adjacent to v lie on the other side of u, we exchange u and v. For every node u, we determine whether a node v exists with such properties. If several appropriate nodes exist the v with highest gain is chosen. Figure 4.1 shows a situation in which nodes u and v are exchanged as the nodes w,x,y adjacent to u are right of v and the nodes s,t adjacent to v are left of u.

Node Insertion

For the node insertion heuristic we test for every node u whether a good new position can be found. To determine such a new position, we apply the idea described above to the old and new position of u. Furthermore, as all nodes between these two positions are moved in case of an insertion, we have to investigate the change of the lengths of their incident edges as well.

Subsequence Reversal

All subsequences of length up to 20 are tested with respect to a possible reversal of all nodes. Again, we apply the change if most of the incident edges of all nodes within the subsequence can be shortened.

Local Enumeration

For the local enumeration heuristic we choose subsequences of length up to 6 and compute the contributions to the objective function value of these segments. Now we proceed from left to right determining the best ordering of each segment.

Kernighan-Lin

The basic scheme of a Kernighan-Lin heuristic is to combine small, simple changes with complex rearrangements of the permutation. Although we allow the simple modifications to worsen the objective function this is not allowed for the complex steps. During a series of complex modifications the step with the maximal improvement is remembered. In the end all modifications are undone and only the memorized best step is performed. We consider a Kernighan-Lin heuristic based on node insertion moves followed by a Kernighan-Lin heuristic that uses 2-exchange moves. For more information and for an insight in a very sophisticated realization of the Kernighan-Lin heuristic see [110].

In Section 5.5.2 we investigate which heuristic is the best choice for our branch-and-cut-and-price algorithm.

4.3 Separation

In this section a short overview of the different separation ideas is given. We distinguish between exact and heuristic separation algorithms, whereas in the first case complete enumeration is used. This was originally done to get an idea of the constraint's effect on the root bound. We then experienced that over 96% of the running time of our branch-and-cut-and-price-algorithm is spend on the solution of the LP. Therefore, we focused on determining the strongest inequalities instead of the development of fast separation procedures.

4.3.1 Exact Separation Algorithms

We give a list of all constraints that are separated with a complete enumeration strategy. The symmetry is exploited such that the real computation time is often not near to the presented worst case complexity.

- The degree-big inequalities (2.5) can be enumerated in $O(n^3)$.

- The same holds for the each-edge-one-distance equations (2.2).

- For the large bipartite and large hypermetric inequalities all active bipartite and hypermetric constraints are considered in view of their possible enlargement. The detailed procedure is described in [97] and takes $O(n^3)$.

- The enumeration of the monotonic inequalities (2.12) has a complexity of $O(n^3)$.

- The single-degree inequalities (2.13) can be enumerated in $O(n^3)$.

- The sparser-star inequalities are star inequalities (1.2), in which we consider the smallest violated star S for a fixed distance k. We obtain this smallest S by the procedure explained in [7]. The advantage compared to ordinary stars is, that this constraint is sparser. The enumeration complexity is $O(n^3)$.

- Special-degree equations (2.4) have an enumeration complexity of $O(n^3)$.

- The star inequalities (1.2) can be enumerated in $O(n^3)$.

- To enumerate the the-longer-the-rarer equations (2.3) we need $O(n^3)$.

- The bipartite and hypermetric inequalities are defined in (1.8) and (1.7). Due to their similarity we separate both inequalities within one enumeration routine. The time complexity for the separation of both constraint types is $O(n^7)$.

- For the separation of bridge inequalities (2.15) to (2.18), we start computing all 1-paths in the current LP solution starting at each node. Then we separate the different types of bridges one after another. Each search for a violated bridge structure is based on the knowledge about all the 1-paths (that do not contain cycles). Although the 1-paths can be computed in $O(n^2)$ and make the enumeration of the bridge structures easier, the maximum complexity in this separation remains to be $O(n^4)$.

- We separate 3-, 4-, 5- and 6-clique inequalities successively, i.e., for every l-clique, violated or not, we test whether it can be increased to a $(l+1)$-clique for $1 \leq l \leq 5$. This gives a complexity of $O(n^7)$, as we do not only need to determine all nodes of the cliques but have to sum over all distances k as well.

- The 3- and 4-cycle-star inequalities displayed (2.29) to (2.30), and the 3- and 4-cycles-with-legs inequalities, see (2.31) to (2.36), are separated by enumerating all 3- and 4-cycles with the corresponding attachments. The highest complexity is $O(n^5)$.

- During the exact enumeration of the triangle inequalities (2.24) the 3-cycle inequalities (1.4) are separated as well. As an alternative, we use the heuristic triangle separation described below and enumerate the 3-cycle separately. In both cases we have a complexity of $O(n^4)$.

- The special-triangle inequalities (2.14) can be enumerated in $O(n^6)$.

- The path-star inequalities, shown in (2.19) to (2.22), are enumerated one after another. The highest complexity is $O(n^4)$ as these constraints are formulated in d-variables and do not contain the y-variables implicitly.

- While enumerating the 3- and 4-prism inequalities (1.6), we separate diamond inequalities (2.28) as well. The reason is that diamonds are substructures of prisms as it can be seen in Figures 1.2-(g) and 2.4. We start with the enumeration of all 3-cycles, $O(n^3)$. Then we determine all possible combinations between two different 3-cycles, which is $O(n^4)$. The same idea is used for the separation of 4-prisms, where the separation needs $O(n^5)$.

- For the subgraph inequalities (2.26) we have to enumerate a star with at most 7 neighbors for each node. Then we add all edges between neighbors of the center node to obtain a complete subgraph G'. The computation of $c_{\text{opt}}(G')$ is an upper bound of the complexity of this separation procedure, hence $O(|G'|)$.

- The same holds for the subset inequalities (2.27). As we observed that nodes with high degrees are often close within an optimal solution, we want to separate on subsets containing of nodes having large degrees. We therefore order the nodes of the graph G by their degrees and then consider the subgraphs D defined by all edges of the $|D|$ nodes, the nodes 2 to $|D|+1$, 3 to $|D|+2$ and so forth. We choose $6 \leq |D| \leq 8$.

- For the subtour inequalities (2.11) we increase the size of the subset S successively: Every time a node is added to S, we test whether the constraint is violated or not. If so, we do not increase the size of S and obtain sparser inequalities than considering the largest S for every given distance k. The time complexity of the algorithm is $O(n^4)$.

- We enumerate 3-, 4- and 5-wheel inequalities (1.5) in $O(n^6)$: For every node l, we test whether a 3-, 4- or/and 5-cycle between l's neighbors exists.

4.3.2 Heuristic Separation Algorithms

In contrast to the exact 3-cycle enumeration, we search for violated cycle inequalities (1.4) of arbitrary length using breadth first search that is started from every node of the graph. The time complexity of this method is $O(n+m) \cdot O(n^2)$.

The idea of a heuristic separation of triangle inequalities (2.24) is described in [45]. Here, the third node l is chosen to be the one with maximal violation. Again this leads to a complexity of $O(n^3)$ but stronger triangle inequalities are chosen.

Besides these two ideas, we combined complete enumeration with a time limit for the separation procedure, which we set to 1 second per call. We applied this heuristic idea to all separation procedures that were remarkably time consuming. These are

- the bipartite and hypermetric inequalities,

- the bridge inequalities,

- the 3-, 4-, 5- and 6-clique inequalities,

- the 3- and 4-cycle-star inequalities,

- the triangle inequalities,

- the special-triangle inequalities,

- the path-star inequalities,

4.3. SEPARATION

- the 3- and 4-prism inequalities,
- the subgraph inequalities,
- the subset inequalities,
- the subtour inequalities, and
- the 3-, 4- and 5-wheel inequalities.

4.3.3 SEPARATION SPEED UP

We will now present different ways to speed up the general separation procedure.

RANDOM AND SMART SEPARATION

The first idea is that the constraints, whose violation has to be tested, are not chosen iteratively but in a random order. Another way is that the separation routines do not start at the same constraint whenever they are called. Instead all constraints tested during the last iterations are skipped and the test of violation starts with the first constraint that has not been considered yet. Only if no violated constraint can be found, we test the first ones as well.

DEEP SEPARATION

A way totally different from the ones described above is based on the idea to work with a modified LP solution $\tilde{d}^*$. First we consider the following convex combination of the original LP solution d^* and the vector d_{feas} corresponding to the best known feasible solution.

$$\tilde{d}^* := (1-\lambda)d^* + \lambda d_{\text{feas}},$$

with $0 \leq \lambda \leq 1$. The idea is to walk along the way between the current LP solution and the best found feasible solution. As the latter is hoped to be near to the optimal solution it might be an advantage to separate a point along the way instead of d^*. A good position on the line has to be determined for the specific problem. We present different choices of λ in Section 5.7.4 and refer to this separation as **deep feasible separation**. This idea is well known by now and was originally presented by G. Reinelt many years ago [109].

For another variant of this idea we choose the vector

$$d_{\text{center}} = \frac{1}{n} \sum_{d_\pi, \pi \in S(n)} d_\pi$$

corresponding to the center of the polytope instead of a feasible solution d_{feas}. In contrast to many other polytopes, such as for example the max cut problem, with a center of $(0.5, \ldots, 0.5)^T$, the situation is more complicated in our case:

$$d_{\text{center}_{ijk}} = \frac{2(n-k)}{n(n-1)} \quad \text{for all } i < j \in V, k = 1, \ldots, n-1.$$

The formula can be understood in the following way: The probability that i takes a certain position is $1/n$. For j only $n-1$ free positions are left, hence the probability for j is $1/(n-1)$. There are $n-k$ possibilities in which i and j have the distance k and due to symmetry of permutations we have a factor of 2. Analog to the first idea we set

$$\tilde{d}^* := (1-\lambda)d^* + \lambda d_{\text{center}},$$

with $0 \leq \lambda \leq 1$, and present different choices of λ in Section 5.7.4. We call this modification of the LP solution the **deep center separation**.

4.3.4 Cut Selection Strategies

Finding an LP solution of our constraint system is relatively time consuming, therefore it is important to keep the system as small as possible. In the following we present two different strategies to choose a small set of effectively violated constraints out of all violated cuts. Both strategies can be applied after every generation of cutting planes.

Rankings

We rank each constraint and order all violated constraints with respect to the ranking. We than add the first $maxConAdd$ constraints to the LP. In Section 5.7.3 we present different values of $maxConAdd$ and compare the usefulness of the various rankings. A good survey about ranking strategies can be found in [111].

Let $A^T d \leq b$ be our constraint system of the MinLA and d^* be an LP solution. Furthermore let $A_i^T d \leq b_i$ denote one row of the constraint system. The first ranking idea is to select the cuts at **random**. Another widely used strategy is to pick the constraint with the maximal amount of violation $A_i^T d^* - b_i$. We call this the **violation ranking**. Two ideas, that were tested for the first time in [111], are the following. First, we select with priority those constraints that have a maximal distance between d^* and the hyperplane $A_i d^* = b_i$. Geometrically speaking this means that the chosen cut has a large cuts-off effect. We call this strategy **distance ranking**. The second idea is to prefer constraints $A_i d^* = b_i$ being as parallel to the objective function as possible. The reason is that we cut of the biggest part of the polytope by choosing such a constraint. This selection strategy is called **angle ranking**.

Variable Disjoint Cut Selection

The small set of chosen constraints can be expected to be more effective when the constraints cover the whole range of variables. That means the constraints should share as less variables as possible. We even want the chosen constraint set to be variable disjoint. Note: This approach is highly dependent on the ordering in which the constraints are separated. Therefore it is reasonable to separate the important constraints before all others. If we use a ranking in addition to the variable disjoint cut selection, we simply sort all violated constraints and then choose all constraints that are variable disjoint to the ones already chosen. The procedure is displayed in Algorithm 10.

Algorithm 10 VariableDisjointCutSelection()

Input: Buffer of violated constraints,
Set of active variables,
Output: Variable disjoint constraint buffer.

1: **for all** violated cuts c **do**
2:   **for all** active vars v **do**
3:     **if** coeff of v in cut c is $> \varepsilon$ **then**
4:       **if** v is marked **then**
5:         Delete c from the buffer of all violated constraints.
6:       **end if**
7:     **end if**
8:   **end for**
9:   Constraint c is variable disjoint to the already chosen cuts.
10:   **for all** active vars v **do**
11:     Mark v.
12:   **end for**
13: **end for**
14: Return modified buffer of violated constraints.

4.4 Improvement Heuristics

During the branch-and-cut-and-price algorithm we want to use the fractional LP solution to construct feasible MinLA solutions. This is done after every cutting plane generation with the hope to improve the heuristic start solution. The idea is that during the algorithm the LP solution gets closer and closer to the optimum. Therefore there must be some information about the optimal solution within the fractional LP solution which we want to encode.

Recalling the definition of our variables ($d_{ijk} = 1$ if and only if the distance of node i and j is k) it seems hard to discover reasonable hints from a fractional LP solution. Nevertheless the following heuristics are based on two observations.

- A node with an adjacent edge of length k is at least k nodes away from one of the borders of the permutation.

- The two end nodes of long edges are near the borders of the permutation.

4.4.1 Distance to the Border Heuristic

We consider the nodes in a random order and determine the maximal k such that $d_{ijk}^* > \varepsilon$, where $\varepsilon = 0.2$. Then the node i is placed in the kth left or k^{th} right position. If both positions have already been taken by other nodes, we choose the first empty position from the left. The procedure is displayed in Algorithm 11.

Algorithm 11 `DistanceToTheBorderHeuristic()`

Input: Current primal variables d^*,
Output: Feasible solution.

1: **for all** nodes $i \in V$ **do**
2:   Set $maxDistance[i] := 0$.
3:   **for all** nodes $j \in V$ adjacent to i **do**
4:     **for all** distances $k = 1, \ldots, n-1$ **do**
5:       **if** $d_{ijk}^* > \varepsilon$ **then**
6:         **if** $k > maxDistance[i]$ **then**
7:           Set $maxDistance[i] := k$.
8:         **end if**
9:       **end if**
10:     **end for**
11:   **end for**
12:   **if** position $maxDistance[i]$ is empty **then**
13:     Place node i at position $maxDistance[i]$. // Place i to the left.
14:   **else**
15:     **if** position $|V| - maxDistance[i]$-1 is empty **then**
16:       Place node i at position $|V| - maxDistance[i]$-1. // Place i to the right.
17:     **else**
18:       Place node i to the first free position from the left.
19:     **end if**
20:   **end if**
21: **end for**
22: Return generated permutation.

4.4.2 Longest Distances Heuristic

Long distances within a permutation can occur only between nodes placed near the borders. Therefore, we developed a heuristic that searches for long distances within the current LP solution and places the two end nodes to the borders of the permutation. With decreasing distance k all edge lengths between the remaining nodes are considered. The end nodes of the next longest edge are placed to the leftmost and rightmost free position in the permutation. Algorithm 12 shows the detailed procedure. Again the threshold ε is set to 0.2.

Algorithm 12 `LongestDistancesHeuristic()`

Input: Current primal variables d^*,
Output: Feasible solution.

1: Determine $i, j \in V$ with $d^*_{ijk} > \varepsilon$ and k maximal.
2: **if** no such nodes i, j can be found **then**
3:    Choose i, j randomly.
4: **end if**
5: Set i to the left most and j to the right most position.
6: **for all** distances $k = n - 1, \ldots, 1$ **do**
7:    **if** position k is empty **then**
8:      **for all** nodes $j \in V$ not placed yet **do**
9:        **if** $d^*_{ijk} > \varepsilon$ **then**
10:          Set j to position k.
11:          Leave the loop.
12:        **end if**
13:      **end for**
14:    **end if**
15: **end for**
16: **if** not all nodes have been placed **then**
17:    Fill empty positions from left to right with all left over nodes.
18: **end if**
19: Return generated permutation.

4.4.3 Edge Lengths Heuristic

This heuristic is motivated by the well known and good working improvement heuristic of the linear ordering problem. For the current LP solution d^*, we sum up the lengths d^*_{ijk} of edges ij incident to i. We sort these sums decreasingly. Following the idea that nodes with a lot of long edges have to be placed near the borders, we start placing the nodes from outside and alternate to the middle. We choose the threshold to be $\varepsilon = 0.2$.

In a second variant, we weight the sum of edge lengths starting at each node with the number of edges. This idea is a specification from the observation above: A huge sum consisting of few edge lengths implicates that the node has to be at the border of the permutation. The described heuristic is displayed in Algorithm 13.

All three improvement heuristics are followed by the same simulated annealing algorithm. Unlike the start heuristic this simulated annealing algorithm starts not with a random permutation but with the solution of the foregone improvement heuristic.

Algorithm 13 EdgeLengthHeuristic()

Input: Current primal variables d^*,
Output: Feasible solution.

1: **for all** nodes $i \in V$ **do**
2:   Set $edgeLengthsSum[i]$:=0.
3:   Set $nAdjNodes[i]$:=0.
4:   **for all** nodes $j \in V$ adjacent to i **do**
5:     Increase $nAdjNodes[i]$ by 1.
6:     **for all** distances $k = 1, \ldots, n-1$ **do**
7:       Set $edgeLengthsSum[i] += k \cdot d^*_{ijk}$.
8:     **end for**
9:   **end for**
10: **end for**
11: Sort $edgeLengthsSum[i]$ decreasing,
    Alternatively sort $edgeLengthsSum[i]/nAdjNodes[i]$.
12: **for all** $i = 0, \ldots, n-1$ **do**
13:   Place node belonging to $edgeLengthsSum[i]$ at position i
14:   Place node belonging to $edgeLengthsSum[i+1]$ at position $n-i$
15:   Increase i by 2.
16: **end for**
17: Return generated permutation.

4.5 Branching

We will now present different branching criteria.

4.5.1 Branch on Variables

Default Branching

The default branching on variables often leads to two subproblems, see Section 0.8.1. The first subproblem defined by $d_{ijk} \leq \lfloor d^*_{ijk} \rfloor$ and the second by $d_{ijk} \geq \lceil d^*_{ijk} \rceil$. Unfortunately, we have no logical implications for the branch in which d_{ijk} is fixed to 0. On the other hand, we can fix a lot in the branch in which d_{ijk} fixed to 1: All $d_{ijk'}$, where $k' \in \{1, \ldots, n-1\} \setminus \{k\}$ can be fixed to 0. This is realized in the ABACUS routine setByLogImp(), see Section 5.2.3. The performance of this branching strategy cannot be expected to be good. This is due to the fact that the tree is highly unbalanced. We therefore come away from a binary branch-and-bound tree and consider other branching strategies.

Set Relative Distances of Two Nodes

A pair of nodes $i < j \in V$ is chosen. The $n-1$ subproblems are defined by the following additional constraints $d_{ijk} = 1$ for $k = 1, \ldots, n-1$. In every node, we have $n-1$ variables set to 0 and the tree is balanced. The longest path from the root to a leaf of the branch-and-bound tree is $\binom{n}{2}$. Algorithm 14 describes the definitions of the subproblems in detail.

4.5.2 Branch on Constraints

In contrast to use variables for branching, we can use constraints as well. We present several different strategies.

Algorithm 14 Set Relative Distances of Two Nodes

Input: Nodes $i, j \in V$,
Output: Subproblems defined by all possible distances dist_{ij} of i, j.

1: **for all** $\text{dist}_{ij} = 1, \ldots, n - 1$ **do**
2:    **for all** active variables v **do**
3:       **if** $v = d_{ij\,\text{dist}_{ij}}$ **then**
4:          Set $d_{ij\,\text{dist}_{ij}}$ to UB 1.
5:       **end if**
6:       **if** $v = d_{ijk}$ with $k \neq \text{dist}_{ij}$ **then**
7:          Set d_{ijk} to LB 0.
8:       **end if**
9:    **end for**
10:   **if** no active variable exists with $v = d_{ij\,\text{dist}_{ij}}$ **then**
11:       Generate variable $d_{ij\,\text{dist}_{ij}}$ and add variable to the LP.
12:       Set its UB to 1.
13:   **end if**
14:   Generate new sub node defined by the above described set variables.
15: **end for**

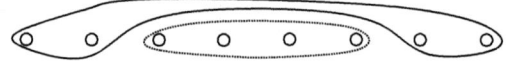

Figure 4.2: Intervals for branching on the deg-big constraint.

Degree-Big

The idea of this branch rule was developed in cooperation with A. Letchford. We thought about branching by assigning nodes to intervals. Because of Equation (2.4) we know that from a fixed node, there are either two incident edges of length k or one incident edge of length $(n - k)$ and one of incident edge of length k. (Incident edges of length not equal to k or $(n - k)$ are not considered here.) We now distinguish between the interval of all positions, from where both incident edges are short, and the interval of all positions from where one incident edge is short and the other long. Figure 4.2 shows both intervals for $n = 8$. All nodes assigned to the interval shown with a dotted line, have exactly two short edges, e. g. of length 2 and 5 or of length 3 and 4. Nodes in the other interval, drawn in a solid line, have exactly one short and one long incident edge, in this example of length 1 and of length 6. The constraint corresponding to this connection between length of incident edges and their number of occurrence is the degree-big constraint (2.5). As we want the branch-and-bound tree to be balanced, we determined that $k = \lfloor n/4 \rfloor$ fulfills this criterion the best. We therefore choose an node i per level in such a way that $\sum_{j \neq i} d_{ij(n-\lfloor n/4 \rfloor)} \leq 1$ has a slack close to 0.5. The two subproblems are defined by the following constraints

$$\sum_{j \neq i} d_{ij(n-\lfloor n/4 \rfloor)} = 0 \quad \text{and}$$

$$\sum_{j \neq i} d_{ij(n-\lfloor n/4 \rfloor)} = 1.$$

Due to the special-degree constraint (2.4) we know: In the first subproblem the node i is forced to lie in the interval with two short incident edges, whereas in the second subproblem the node i has one long and one short incident edge.

4.5. Branching

Triangles

Again a triple $i < j < l \in V$ of nodes is chosen in every level. We define three subproblems that correspond to the different orderings in which these nodes can occur. If node j is placed between node i and l the triangle inequality $y_{ij} + y_{jl} \leq y_{il}$ is satisfied with equality. If node i lies between j and l, we have $y_{ij} + y_{il} = y_{jl}$. In the last case, in which l is between i and j, $y_{il} + y_{lj} = y_{ij}$ holds. Hence, each subproblem is defined by one of the three triangle equalities corresponding to i, j, l. This idea was suggested by A. Letchford [96].

3-Cycles with Odd Right Hand Side

This branch rule is based on the fact, that no cycle can have a linear arrangement with an odd objective function value. We take advantage of this fact in the following way: For every level, we identify a 3-cycle (C, E) with odd right hand side and define two subproblems by the constraints

$$\sum_{ij \in E} y_{ij} \geq \text{rhs} + 1 \quad \text{and}$$

$$\sum_{ij \in E} y_{ij} \leq \text{rhs} - 1.$$

Subgraphs

Let $G' = (V, E)$ be a star with all edges between the neighbors of the center nodes. We compute the MinLA solution $c_{\text{opt}}(G')$ and a trivial upper bound UB for G'. The last one is based on the idea of the so-called "edge lower bound", see Section 1.5.1. Now, $UB - c_{\text{opt}}(G')$ subproblems are define by the additional constraints

$$\sum_{ij \in E} y_{ij} = c_{\text{opt}}(G'),$$

$$\sum_{ij \in E} y_{ij} = c_{\text{opt}}(G') + 1,$$

$$\vdots$$

$$\sum_{ij \in E} y_{ij} = UB.$$

To assure that the integral solution can be found within the branch-and-bound tree, we have to start variable branching when no branch constraint can be found anymore.

4.5.3 Branch on Variables and Constraints

We will now combine the two general branching approaches in the following way: For one single level of the branch-and-bound tree, we define some subproblem by constraints and others by variables.

3-Cycles with Odd Right Hand Side

From our observations, we know that even for large instances a lot of small cycles are placed optimally in the optimal solution for the whole instance. Therefore, we develop the idea described in Section 4.5.2 a bit further. Again, we search for a 3-cycle $G' = (C, E)$ with odd right hand side. Now we define four subproblems in the following way. The first subproblem is generated by the additional constraint

$$\sum_{ij \in E} y_{ij} \geq c_{\text{opt}}(G') + 2.$$

The subproblem corresponding to $\sum_{ij \in E} y_{ij} = c_{\text{opt}}(G')$ is splitted up: Every variable setting that fulfills this constraint with equality is realized in another subproblem. I. e., the second subproblem is defined by the optimal arrangement of the 3-cycle i,j,l in this ordering of the nodes.

$$d_{ij1} = 1, d_{jl1} = 1, d_{il2} = 1, \text{ and}$$
$$d_{ijk} = 0, d_{jlk} = 0 \text{ for all } k \in \{1,\ldots,n-1\} \smallsetminus \{1\}, \text{ and}$$
$$d_{ilk} = 0 \text{ for all } k \in \{1,\ldots,n-1\} \smallsetminus \{2\}.$$

For the third subproblem all variables are set in such a way, that the nodes i,l,j are in consecutive order.

$$d_{il1} = 1, d_{lj1} = 1, d_{ij2} = 1, \text{ and}$$
$$d_{ilk} = 0, d_{ljk} = 0 \text{ for all } k \in \{1,\ldots,n-1\} \smallsetminus \{1\}, \text{ and}$$
$$d_{ijk} = 0 \text{ for all } k \in \{1,\ldots,n-1\} \smallsetminus \{2\}.$$

In the fourth subproblem the same is realized for the consecutive arrangement of the nodes j,i,l.

4.5.4 Choice of Branch Variable/Constraint

The choice of the branching variables and constraint is essential for the successful application of the branch rule. We therefore present several criteria for suitable choices.

Choice of Suitable Branching Variables

For the choice of two branching variables, we have the following criteria. Choose $i,j \in V$ such that

- The key idea of this choice is to balance the branch-and-bound-tree. If a distance y^*_{ij} consists of only one non-zero d^*_{ijk} value, the distance of i and j seems to be quite sure. Therefore, the tree would probably be very unbalanced if we choose this pair of nodes. Therefore, choose i and j such that the distances y^*_{ij} consist of more than one non-zero value of the corresponding d-variables.

- We refine the idea and choose that pair i,j that has the maximal number of non-zero values d^*_{ijk} greater than 0.3.

In case of three suitable branching variables per level, we suggest the following criteria.

- Compute the slacks of the three corresponding triangle constraints that are going to define the subproblems. Choose $i,j,l \in V$ for which the minimal slack is maximal. That means in this branching step the minimal effect in all three subproblems is maximal, which can be seen as a way of balancing the branch-and-bound tree.

- Again consider the tree slacks. Now choose that triple, for which the sum of all three slack is maximal. Here, we do not focus on an effect in all subproblems but want to maximize the overall effect. A drawback is that one subproblem might have a huge slack while the other have a very small slack. Hence the tree might not be balanced.

Choice of Suitable Branching Structure G'

It is important to assure that a different branch constraint is chosen at every level of the tree. Since we want to branch in such a way that the tree is balanced, we suggest the following procedure. We choose the constraint whose current labeling is far away from the optimal arrangement. Algorithm 15 explains the choice of G' in detail.

4.5. Branching

Algorithm 15 Choice of Suitable Branching Structure G'

Input: Input graph $G = (V,E)$,
Current LP solution d^*,
Output: Branching Structure G'.

1: Set $max := 0$.
2: Set $best := 0$.
3: Initialize G'_{best} with the empty set $\emptyset$.
4: **for all** nodes $l \in V$ **do**
5:     Compute subgraph $G' = (V', E')$ with center node l.
6:     Set $best := \sum_{ij \in E'} y^*_{ij} - c_{opt}(G')$.
7:     **if** $best > max$ **then**
8:         Set $G'_{best} := G$.
9:         Set $max := best$.
10:     **end if**
11: **end for**
12: Return G'_{best}.

CHAPTER 5

COMPUTATIONAL RESULTS

In this chapter we describe the impact of different settings of the branch-and-cut-and-price algorithm. We start with some implementation details and a presentation of the relevance of the binary distance modeling of the MinLA. In Section 5.4 we describe the tests for the complete model before we focus on pricing aspects in Section 5.5. Here, we are particularly interested in the choice and number of the start variables. Furthermore, we compare different numbers of priced-in variables and consider the frequency of additional pricing steps. In Section 5.6 we turn our attention to the strength of each constraint type and on finding the best combination of constraint types. This is followed by several ideas to improve the separation procedures in Section 5.7. The usefulness of different improvement heuristics is discussed in Section 5.8. Furthermore, in Section 5.9 we outline the various branching ideas and present their impact on the computational results. In Section 5.10 the corresponding sparse problem formulation of the MinLA is investigated. We display the effect of additional variables and the shortest strengthening. The chapter concludes with a general comparison of lower bounds for the MinLA in Section 5.11.

5.1 TEST PROBLEM INSTANCES

We use two different collections of test graphs, which are well-known in the context of the MinLA problem. The first one is the classical benchmark collection for the MinLA which was introduced by Petit i Silvestre in [2]. As most of the instances have a large number of nodes, we consider the instances with $n \leq 180$. "The Harwell-Boeing Sparse Matrix Collection is a set of standard test matrices arising from problems in linear systems, least squares, and Eigenvalue calculations from a wide variety of scientific and engineering disciplines. The problems range from small matrices, used as counter-examples to hypotheses in sparse matrix research, to large test cases arising in applications. The collection was originally developed by Iain Duff, Roger Grimes, and John Lewis." [112] We consider all instances where the number of nodes is $\leq \text{argmin}_n\{\text{gd95c},\ldots,\text{gd96d}\}$. In Table 5.1 we summarize some information about the chosen instances. In column two and three, the number of nodes and edges are given. The density of the graphs is displayed in column four. The trivial lower bound LB_{trivial} is obtained by the degree method, see Section 1.5.1. The upper bound UB in column six is computed with the multi-start local search routine described in Section 4.2.2.

Table 5.1: Properties of the test instances.

Name	n	m	Density	$LB_{trivial}$	UB
bcspwr01	39	46	0.01	58	109
bcspwr02	49	59	0.05	81	173
bcsstk01	48	176	0.16	431	1152
can_24	24	68	0.25	142	210
can_61	61	248	0.14	746	1137
can_62	62	78	0.04	101	222
curtis54	54	124	0.9	240	512
ibm32	32	90	0.18	189	493
impcol_b	59	281	0.16	970	2358
pores_1	30	103	0.24	239	383
will57	57	127	0.8	248	352
gd95c	62	144	0.08	292	506
gd96b	111	193	0.03	702	1416
gd96c	65	125	0.06	191	519
gd96d	180	228	0.01	595	2289

5.2 Computational Setup and Details of the Implementation

The algorithm was implemented in C/C++ and embedded in the branch-and-cut framework ABACUS ("A Branch-And-CUt System", see [35]). The algorithm was run on a 2× Xeon CPU with 2.5GHz, 2 × 6MB Cache and 2GB RAM under Debian GNU/Linux 4.0 using CPLEX 8.1 [26]. All running times are given in seconds.

5.2.1 Memory Management of Variables due to Branch-and-Price

Within a branch-and-cut-and-price algorithm the set of active variables is expected to be different in every node of the branch-and-bound tree. Therefore, a lot of information has to be saved in every sub node instead of a central information collection in the master of the optimization. Especially the variables themselves and the LP solution have to be saved in every sub node. As the algorithm shall be useful for large instances, we want to avoid a memory structure of n^3. To make a compromise between running time and memory, we use an $n \times n$ array combined with a list of all variables d_{ijk} that exist for every pair i,j. Depending on the size of the start set of variables, we observed that only up to four variables for each pair i,j are priced-in during the algorithm. Therefore, the list attached to each array entry is very short and the access and running times are acceptable.

5.2.2 Central Saved Adjacency List

Besides the above described memory management, there is one important information that does not depend on the active variables and can therefore be stored in the master of optimization: The adjacency list of each node is computed once in the master of the problem. It can then be used during the whole algorithm. This reduces the access of all adjacent nodes for i from $O(n)$ to $O(\deg(i))$.

5.2.3 SetByLogImp()

It was mentioned in the beginning of the last chapter that the variables used in this modeling of the MinLA are beneficial with respect to branching. This fact is exploited by the function setByLogImp(). The time of calling of the routine setByLogImp() within the procedure of the branch-and-cut-and-price algorithm can be found in Figure 2. In this function, we test whether one variable d_{ijk} is fixed to one. If this is the case, all $n - 2$ variables $d_{ijk'}$ where $k' \in \{1, \ldots, n-1\} \setminus \{k\}$ can be fixed to 0. If in addition $k \geq \lfloor (n-1/2) \rfloor + 1$, the variables d_{ilk} and d_{jlk} can be fixed to 0 for all $l \in V \setminus \{i, j\}$. These two logical implications directly correspond to the each-edge-one-distance equations and degree-big constraints.

5.3 Justification of the 0/1 Model

Before we start with the detailed test of different settings of the algorithm, we present a comparison between the distance model and the binary distance modeling approach. To be precise, we show how the lower bounds change if we use the d-variables and the constraints in addition to the y-variables and their constraints. In the first three columns of Table 5.2, information about the instances is given. In column four and five, the lower bounds, LBs, of both variants are presented. This is followed by the gap closure between the LB and the UB. (For the definition of the gap closure see Section 0.8.3 on page 13.) In the last two columns, the time with and without the d-variables is presented. Detailed information about the general setting that is used to obtain the results can be found in Section 5.7.4.

We have to be careful interpreting the results of Table 5.2. This is because the LB is only reliable if the root bound is reached within the time limit. (The LB obtained just before the branching starts is called **root bound** and abbreviated with **RB**.) In all other cases, the LB might be decreased due to the pricing-in of variables, please read Section 5.5.1 on page 95 for an explanation. Therefore, the important rows of Table 5.2 are the ones for can_24, ibm32, and pores_1. It can be seen that the gap closure lies between 42% and 53%. The "fine structure" of the d-variables is therefore indeed important to specify the properties of a linear arrangement. Instances where the graph closure is marked with "-" have a LB above the UB, that is why no gap closure is given for these instances.

Table 5.2: Improvement of binary distance model compared to the integral distance modeling approach. (Time limit is 86 400 seconds. LBs marked with "*" are above the UB because of pricing, see Section 5.5.1 on page 95 for details.)

Instance			LB			Time in seconds	
Name	n	m	No d-vars	With d-vars	Gap cl.	No d-vars	With d-vars
bcspwr01	39	46	87.59	103.91	76.23%	179	limit
bcspwr02	49	59	137.02	172.93	99.81%	662	limit
bcsstk01	48	176	896.23	1151.62	99.72%	853	limit
can_24	24	68	196.76	203.86	53.63%	23	1080
can_61	61	248	1003.8	1137	100%	2217	limit
can_62	62	78	173.98	221.99	99.98%	8903	limit
curtis54	54	124	354.37	511.91	99.94%	4870	limit
ibm32	32	90	439.02	462.36	43.24%	67	30677
impcol_b	59	281	1277.98	2357.81	99.98%	8706	limit
pores_1	30	103	327.47	351.02	42.41%	55	15340
will57	57	127	236.27	352	100%	9705	limit
gd95c	62	144	281.42	559*	-	350	limit
gd96b	111	193	382.2	1619.73*	-	3362	limit
gd96c	65	125	244.05	525.31*	-	21927	limit
gd96d	180	228	816.12	2494.67*	-	limit	limit

5.4 Complete 0/1 Model

In the first place, we are interested in the choice of variables, i.e., whether it is better to use the d- and y-variables explicitly or not. The second basic decision is the solution method of the linear programming relaxations within our branch-and-cut-and-price-algorithm. These two settings are tested without pricing, hence, we have all variables in our system. Only the IP inequalities (2.2) to (2.7) are separated and no sophisticated general separation strategy or cut selection procedure is applied. In each iteration at most 250 cuts are generated of each type. As a limit for the running time we chose 24 hours (86 400 seconds).

5.4.1 Explicit Use of y-Variables

The close relationship between the d- and the y-variables raises the question whether it is better to use the y-variable explicitly or not. In theory, this makes no difference for the computation, as Equation (2.1) allows an internal replacement of the y-variables. We investigated this aspect in practice and present the result in Table 5.3. The first three columns show the name of the test instances, their number of nodes n, and the number of edges m. In the following columns, we compare the explicit setting with the implicit use of y-variables. In column four and five, the number of linear programs, for short #LPs, is shown. The running times are displayed in the next two columns. Columns eight and nine show the LB obtained with both settings. The last column presents the gap closure of the two settings.

In all cases in which the RB can be computed, it is equal for both settings (which has to be the case). For all these instances the RB is obtained in much less time with the implicit than wit the explicit setting.

If the time limit is reached before the RB is reached, a difference in the LB is possible—and such a difference is indeed observed. Within the same time a better LB can be obtained using the y-variables implicitly. We therefore conclude that the advantage of sparse formulated y-constraints is not that important and therefore use the y-variables implicitly from now on.

5.4.2 LP Solver Settings

We now investigate different LP solving strategies and present the results in Table 5.4. Again, the first three columns show information about the instances, where the following columns present the LB and the time to compute this bound for all three settings. In all cases, the barrier method is used to solve the initial LP. The default setting tries to choose between the dual and primal simplex in such a way that phase 1 of the simplex method is not required. We compare this setting with the exclusive use of the dual and primal simplex method. For the smallest instances `can_24`, `ibm32`, and `pores_1`, the RB was computed within the time limit. For these instances, the dual simplex is the fastest setting. This is, however, not the case for the next smallest instance `bcspwr01`. The RB was computed as well but nevertheless the primal setting is the fastest. For all larger instances we can see that the dual simplex is much slower than the primal. It is sometimes even worse than the default setting, see `can_62`, `will57`, and `gd95c`, whereas it computes more or less the same LB for all other cases. When we had a closer look at the running times, we could see that for all instances with more than 40 nodes the dual simplex needed up to 20 hours to solve one single LP. We therefore conjecture that the LPs are dual degenerated. If we consider `bcssstk01`, `gd96c`, and `gd96d`, one can see that the primal simplex computes a significantly higher LB within the same time. We therefore use the primal simplex exclusively from now on.

Table 5.3: Explicit versus implicit use of y-variables.

Instance			#LPs		Time in sec.		LB		
Name	n	m	Expl.	Impl.	Expl.	Impl.	Expl.	Impl.	Gap cl.
bcspwr01	39	46	27	21	15603	4460	88.9	88.9	0%
bcspwr02	49	59	34	30	79651	53785	141.67	141.67	0%
bcsstk01	48	176	21	38	limit	45603	604.3	964.03	65.68%
can_24	24	68	10	7	87	54	200.63	200.63	0%
can_61	61	248	10	17	limit	limit	824.58	885.53	19.51%
can_62	62	78	11	17	limit	limit	111.26	127.43	14.6%
curtis54	54	124	27	27	limit	limit	352.95	366.97	8.81%
ibm32	32	90	16	13	2211	708	443.01	443.01	0%
impcol_b	59	281	10	22	limit	limit	1038.04	1194.45	11.85%
pores_1	30	103	12	9	818	281	328.37	328.37	0%
will57	57	127	13	23	limit	limit	268.87	302.68	40.67%
gd95c	62	144	12	20	limit	limit	379.81	405.294	20.19%
gd96b	111	193	11	15	limit	limit	1203.14	1226.2	10.83%
gd96c	65	125	10	13	limit	limit	218.27	266.845	16.15%
gd96d	180	228	18	19	limit	limit	675.865	685.144	0.58%

Table 5.4: Comparison of different LP solver settings.

Instance			Default		Only Primal Simplex		Only Dual Simplex	
Name	n	m	LB	Time	LB	Time	LB	Time
bcspwr01	39	46	88.9	12519	88.9	10065	88.90	36934
bcspwr02	49	59	112.97	limit	141.67	limit	111.98	limit
bcsstk01	48	176	657.69	limit	919.74	limit	657.68	limit
can_24	24	68	200.63	69	200.63	66	200.63	57
can_61	61	248	808.27	limit	861.54	limit	808.27	limit
can_62	62	78	127.54	limit	128.07	limit	97.63	limit
curtis54	54	124	353.1	limit	356.16	limit	354.37	limit
ibm32	32	90	443.01	1248	443.01	1740	443.01	1170
impcol_b	59	281	1154.88	limit	1217	limit	1154.88	limit
pores_1	30	103	328.37	274	328.37	284	328.74	271
will57	57	127	262.3	limit	303.97	limit	253.34	limit
gd95c	62	144	406.61	limit	405.4	limit	350.74	limit
gd96b	111	193	1143.47	limit	1226.2	limit	1143.47	limit
gd96c	65	125	201.68	limit	268.25	limit	214.03	limit
gd96d	180	228	521.94	limit	696.23	limit	521.94	limit

5.5 Pricing for the 0/1 Model

We now turn from the setting with all variables d_{ijk} for $i < j \in V, k = 1, \ldots, n-1$ to the pricing setting. An overview of the chosen parameters is presented in the following.

- Only the IP constraints (2.2) to (2.7) are separated.
- Triangle inequalities are separated heuristically.
- No sophisticated general separation strategy or cut selection procedure is applied.
- At most 250 cuts per iteration are generated of each constraint type.
- All generated cuts are added to the LP.
- Implicit use of y-variables.
- Primal simplex is used for solving the LP relaxation.
- A feasible start solution π is computed with a heuristic.
- Start variables are all variables $d_{ijk'}$ with $|\pi(i) - \pi(j)| = k'$ for $|k' - k| \le 2$.
- At most 200 variables were priced-in per call.
- No sophisticated pricing strategy is used.
- No additional pricing steps were performed.
- Limit for the running time is 24 hours (86 400 seconds).

Before we start with the tests, we explain a phenomenon that occurs in a branch-and-cut-and-price algorithm.

5.5.1 Lower Bound Decrease Due to Pricing

For a minimization problem, the lower bound rises during the iterations of a branch-and-cut-algorithm. In a branch-and-cut-and-price-algorithm, the situation is more complicated: With every pricing step, the LB decreases by a small percentage. It is even possible that a LB is temporarily above the UB. In this case, it is obvious that the LB is not yet reliable, but even if the LB remains below the UB, we cannot be sure whether the LB is going to decrease or not. We therefore have to be very careful interpreting the LB. Only if the RB is reached, we can make any statement about the LB. Due to this phenomenon, we tested the two very fundamental questions about our algorithm without pricing to have more instances that can be considered for the decision. Figure 5.1 shows a typical behavior of the LB in the root node, displayed for the test instance `pores_1`. The figure shows that the highest LB is reached before the first variables are priced-in. All in all, the LB decreases by around 14%. For the test instance `ibm32`, it is about 10%, for `can_24` around 5%, and for `bcspwr01` the decrease is 16%. No relationship between the decrease of the LB and the density of the instances, nor their number of nodes or number of edges can be found.

5.5.2 Start Sets of Variables

We now turn our attention to the tests of our pricing settings. The first important question is the selection of start variables. As we want to have as few variables in our system as possible, it is essential to start with the right variables. In Table 5.5 we present four different possibilities to choose the start variables. In the first column, the test instance is given. In column two and three, the LB and running time for the identity start solution is shown. The next two columns display the LB and time for a randomly chosen start solution. This is followed by the information about the solutions obtained by the simulated annealing and the multi-start local search approach.

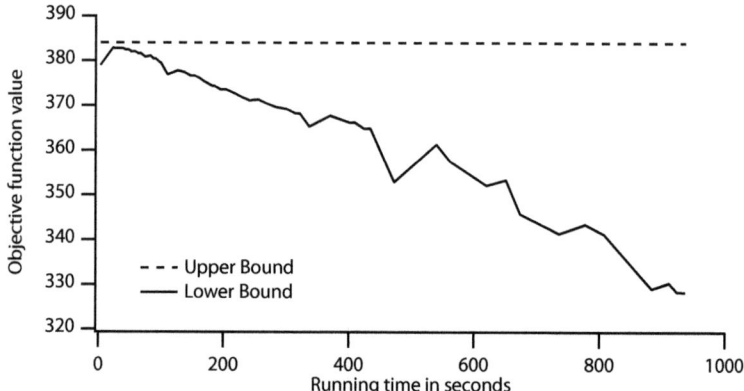

Figure 5.1: Behavior of the LB in a branch-and-cut-and-price algorithm.

Table 5.5: Comparison of different start sets of variables. (LBs marked with "*" are above the UB because of pricing, see Section 5.5.1 on the previous page for details.)

	Identity		Random		SA		MLS	
Instance	LB	Time	LB	Time	LB	Time	LB	Time
bcspwr01	102.88	limit	194.07*	limit	88.9	24920	88.9	29719
bcspwr02	728.4*	limit	764.56*	limit	209.15*	limit	170.97	limit
bcsstk01	2200.98*	limit	2151.54*	limit	1218.96*	limit	1150.92	limit
can_24	200.63	1361	200.63	1042	200.63	408	200.63	578
can_61	4850*	limit	4787.44*	limit	2049*	limit	1137	limit
can_62	522*	limit	1459.19*	limit	306.97*	limit	222	limit
curtis54	1912.23*	limit	1716.13*	limit	529*	limit	512	limit
ibm32	443.01	13492	443.01	8557	443.01	6053	433.01	5049
impcol_b	5318.59*	limit	5362.66*	limit	2083.38	limit	2355.45	limit
pores_1	328.37	9322	328.37	5517	328.37	5241	328.37	3224
will57	2026.42*	limit	2060.8*	limit	437*	limit	352	limit
gd95c	951*	limit	2558.27*	limit	770*	limit	559*	limit
gd96b	9800*	limit	6870*	limit	2545*	limit	1832*	limit
gd96c	2479.25*	limit	2616.07*	limit	565*	limit	526*	limit
gd96d	11537*	limit	13276*	limit	8337*	limit	3293*	limit

5.5. Pricing for the 0/1 Model

Table 5.6: Comparison of different sizes of the start set of variables.

Instance	$\binom{n}{2}$		$\binom{n}{2}+2$		$\binom{n}{2}+4$		$\binom{n}{2}+6$	
	LB	Time	LB	Time	LB	Time	LB	Time
bcspwr01	88.9	29719	88.9	21519	88.9	45418	88.9	29465
can_24	200.63	578	200.63	268	200.63	176	200.63	161
ibm32	433.01	5049	443.01	4699	443.01	6787	443.01	5812
pores_1	328.37	3224	328.37	2807	328.37	3304	328.37	303.8

The results show that only for the instances bcspwr01, can_24, ibm32, and pores_1, the RB was computed within the time limit. As we cannot make any statements about the LB for the other instances, a comparison of the different start sets of variables can only be made for these four instances. We will therefore continue with the tests only with the four instances mentioned above.

For the instances ibm32 and pores_1, the time decreases from left to right, i.e., using the identity as a start solution is the worst choice, whereas the multi-start local search routine turns out to be the best. In case of can_24 and bcspwr01, the time can be improved significantly using the simulated annealing (SA) or the multi-start local search routine (MLS). Simulated annealing is even a bit faster than the MLS. The median of the percental improvement of the simulated annealing algorithm compared to the multi-start local search routine (76%) is very close to the median of the percental improvement of MLSR compared to SA (72%). We therefore choose the MLSR as we know from other tests that it computes a far better heuristic solution than the SA.

5.5.3 Size of the Variable's Start Set

We continue to test different sizes for the start set of variables. The idea is that the start solution might be near to the optimum (this conjecture is encouraged by Caprara et al. [7]). Hence, variables similar to the ones of the start solution itself might be of great help. As we explained above, we do not need to consider all test instances but only those for which the RB was computed within the time limit. In Table 5.6 we present four different settings. In column two and three, we show the smallest possible set of start variables. In the following columns, we add more and more variables close to this solution, see Section 4.1 for an explanation.

The results in Table 5.6 show the following: For the test instance can_24, the setting $\binom{n}{2}+6$ is the best before $\binom{n}{2}+4$. In all three other cases, the setting $\binom{n}{2}+2$ is the best. We therefore choose $\binom{n}{2}+2$ from now on.

5.5.4 Maximal Number of Priced-in Variables per Call

Having sorted out the best set of variables to start with, we consider the pricing-in of variables. In Table 5.7 different numbers of maximal priced-in variables per call are presented. As we want to price-in as few variables as possible, we start with a small number in column two and three, and test up to 200 in column eight and nine.

One can see that the fastest setting for all instances is to generate and add at most 200 variables per pricing step.

5.5.5 Random versus Smart Pricing

We will now compare different general pricing strategies. The ideas described in Section 4.3.3 are adopted to the pricing-in of variables. In column two and three of Table 5.8, no general pricing strategy is applied, whereas in the following columns, the strategies are applied individually and combined.

The results of Table 5.8 show the following: For bcspwr01 and pores_1, using both strategies is fastest. In case of can_24, the random strategy is the best, and for ibm32, the smart strategy turns out to be the fastest. We

Table 5.7: Comparison of different maximal numbers of priced-in variables per call.

Instance	10		50		100		200	
	LB	Time	LB	Time	LB	Time	LB	Time
bcspwr01	88.9	21519	88.9	31904	88.9	21607	88.9	17370
can_24	200.63	268	200.63	108	200.63	84	200.63	64
ibm32	443.01	4699	443.01	2355	443.01	2282	443.01	1295
pores_1	328.37	2807	328.37	1484	328.61	1034	328.37	936

Table 5.8: Comparison of different pricing strategies.

Instance	Default		Smart		Random		Both	
	LB	Time	LB	Time	LB	Time	LB	Time
bcspwr01	88.9	17370	88.9	11541	88.9	17372	88.9	9647
can_24	200.63	64	200.63	66	200.63	61	200.63	62
ibm32	443.01	1295	443.01	786	443.01	1276	443.01	952
pores_1	328.37	936	328.37	531	328.37	540	328.41	494

therefore choose to use both strategies from now on.

5.5.6 Additional Pricing Steps

It can sometimes be of help to perform additional pricing steps. This is because important variables tend in to be priced-in early during the algorithm. We therefore test additional pricing steps after every second solved LP compared to additional pricing steps after every fifth solved LP compared to no additional pricing steps at all. Table 5.9 shows that in all cases, additional pricing steps after every fifth solved LP is the fastest setting.

Table 5.9: Comparison of different frequencies of additional pricing steps.

Instance			No add. steps		After every 5th LP		After every 2nd LP	
Name	n	m	LB	Time	LB	Time	LB	Time
bcspwr01	39	46	88.9	9647	88.9	3837	88.9	6685
can_24	24	68	200.63	62	200.63	55	200.63	66
ibm32	32	90	443.01	952	443.01	774	443.01	1937
pores_1	30	103	328.41	494	328.37	409	328.37	485

5.6 Constraints

We now consider the impact of all the inequalities introduced in Chapter 2. In the following, we use the substructure of a graph to denote its corresponding inequality. For example we write cliques instead of clique inequalities. As a reference LB, we use the best results so far, presented in columns six and seven of Table 5.9.

5.6.1 Strength of Constraint Types

In a first step, we test all inequalities separately and show their influence on the LB in Tables 5.10 to 5.12. For every constraint, we show the LB obtained by the constraint individually and the running time.

Table 5.10: Strength of constraint types (A–H).

Instance	Bipartites		Bridges		Cliques		Cycles		Diamonds		Hypermetrics	
	LB	Time	LB	Time	LB	Time	LB	Time	LB	Time	LB	Time
bcspwr01	89.29	14218	88.9	9406	88.9	11538	89.37	11663	89.3	10970	88.9	25482
bcspwr02	155.48	limit	143.71	limit	145.26	limit	141.89	limit	142.83	limit	165	limit
bcsstk01	1075.34	limit	1010.51	limit	1016.04	limit	1103.52	limit	994.88	limit	1083.89	limit
can_24	201.8	76	200.63	68	201.25	62	200.63	92	201.61	47	200.64	553
can_61	1137	limit	1136.94	limit	1137	limit	1136.86	limit	1137.99	limit	1136.99	limit
can_62	221.67	limit	221.4	limit	221.35	limit	221.69	limit	221.67	limit	221.65	limit
curtis54	503.9	limit	504.48	limit	509.19	limit	510.5	limit	504.18	limit	509.02	limit
ibm32	446.73	2325	443.01	2021	443.01	2555	443.01	3651	443.01	2480	446.06	7114
impcol_b	2355.14	limit	2355.39	limit	2349.3	limit	2352.5	limit	2352.44	limit	2352.35	limit
pores_1	349.85	1505	328.37	974	328.37	845	328.37	1387	328.37	1292	330.19	5290
will157	351.32	limit	350.8	limit	350.91	limit	350.14	limit	352.08	limit	351.16	limit

5.6. Constraints

Table 5.11: Strength of constraint types (I-Sp).

Instance	Monotonics		Path-Stars		Prisms		Single-Degrees		Sparser-Stars		Special-Triangles	
	LB	Time	LB	Time	LB	Time	LB	Time	LB	Time	LB	Time
bcspwr01	88.9	26096	88.9	10032	88.9	10874	88.92	11355	88.9	8953	88.9	112479
bcspwr02	170.17	limit	171.94	limit	141.67	limit	141.67	limit	151.51	limit	150.38	limit
bcsstk01	1132.67	limit	1033.77	limit	972.69	limit	964.04	limit	1035.38	limit	1036.45	limit
can_24	200.78	101	200.63	80	200.63	80	200.63	76	200.63	70	210	limit
can_61	1137	limit	1136.71	limit	1136.45	limit	1136.32	limit	1136.96	limit	1137	limit
can_62	221.5	limit	221.5	limit	221.3	limit	221.55	limit	221.13	limit	221.51	limit
curtis54	509.9	limit	511.91	limit	506.67	limit	501.47	limit	506.09	limit	506.58	limit
ibm32	443.02	3859	443.01	2143	444.02	1613	443.01	2588	443.01	1871	443.01	3226
impcol_b	2354.41	limit	2353.03	limit	2357.64	limit	2354.73	limit	2352.02	limit	2353.86	limit
pores_1	329.37	2093	328.37	1022	339.15	866	328.37	603	328.37	814	328.37	1946
will57	351.13	limit	350.86	limit	351.03	limit	351.04	limit	351.41	limit	352	limit

Table 5.12: Strength of constraint types (St–Z).

Instance	Stars		Subgraphs		Subsets		Subtours		Wheels	
	LB	Time	LB	Time	LB	Time	LB	Time	LB	Time
bcspwr01	88.9	11141	88.9	9120	88.9	11253	88.9	10945	88.9	10877
bcspwr02	141.8	limit	155.75	limit	144.85	limit	141.99	limit	141.85	limit
bcsstk01	998.56	limit	1015.62	limit	996.86	limit	1035.54	limit	999.8	limit
can_24	200.63	64	201.52	65	200.63	70	200.63	59	201.5	71
can_61	1136.84	limit	1136.98	limit	1136.94	limit	1136.73	limit	1136.98	limit
can_62	221.17	limit	221.65	limit	221.13	limit	221.51	limit	221.39	limit
curtis54	503.65	limit	505.59	limit	510.19	limit	507.38	limit	510.91	limit
ibm32	443.09	2008	443.01	2379	443.01	2890	443.03	2374	443.01	2255
impcol_b	2351.88	limit	2355.41	limit	2351.04	limit	2355.86	limit	2353.23	limit
pores_1	328.37	761	328.37	818	328.37	791	328.41	1119	328.37	1206
will157	351.34	limit	351.14	limit	351.86	limit	351.61	limit	350.64	limit

5.6. Constraints

Table 5.13: Constraints chosen by evaluation I.

Name of constraint	#total LB improvements
Bipartites	4
Hypermetrics	3
Monotonics	3
Diamonds	2
Prisms	2
Subtours	2
Wheels	1
Subgraphs	1
Stars	1
Single-Degrees	1
Cycles	1
Cliques	1

We use the results of Tables 5.10 to 5.12 to determine the strongest constraint types. Two criteria shall help to evaluate the results.

- How many times does a constraint improve the LB?
- How large is the improvement?

Interpreting the LB results, we have to keep in mind that for most instances, the LB might decrease during the pricing of variables, recall the phenomenon described in Section 5.5.1. We therefore perform two separate evaluations of the LB results of Tables 5.10 to 5.12. The first considers only the instances `bcspwr01`, `can_24`, `ibm32`, and `pores_1` for which the RB can be computed within the time limit. For these instances, we know that the LB is not going to decrease anymore. Hence, we can interpret the effect of the constraint types right away.

In a second step, we will evaluate the LB of *all* instances, including the four smallest ones. We perform this second analysis as we do not make such an important decision as the selection of cuts to be separated only for very small instances. Many constraint types can unfold their impact on bigger graphs only. Furthermore, we observed that, no matter which constraint types we separated, the LB decreased by the same relative amount. We therefore assume that if a LB_A is higher than a LB_B, the RB_A will be higher then the RB_B as well. We will therefore evaluate the LB of all instances as well, although the results are not as reliable as the results of the four smallest instances. (As the LB of the graph drawing instances `gd95c` to `gd96d` are far above the UB, we do not consider them at all. Their LB has to decrease so drastically, that a comparison of the current LB seems to be useless.)

We start with the evaluation of the small instances.

Evaluation I

We will now consider the results of Tables 5.10 to 5.12 for the four smallest instances mentioned above. We displayed the results in Figure 5.2 and Figure 5.3.

We count how many times a LB was improved by the constraint and rank the constraints with at least one improvement in Table 5.13. Note that we do not consider the amount of the improvements but count all total LB improvements.

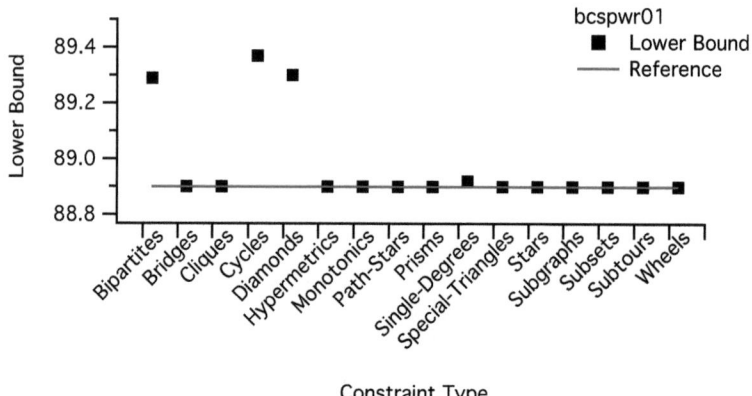

(a) Strength of the constraints for bcspwr01.

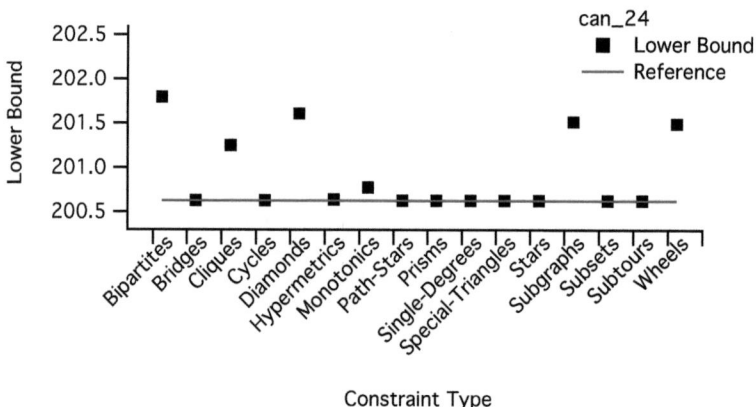

(b) Strength of the constraints for can_24.

Figure 5.2: Strength of all types of constraints for the test instances bcspwr01 and can_24.

5.6. Constraints

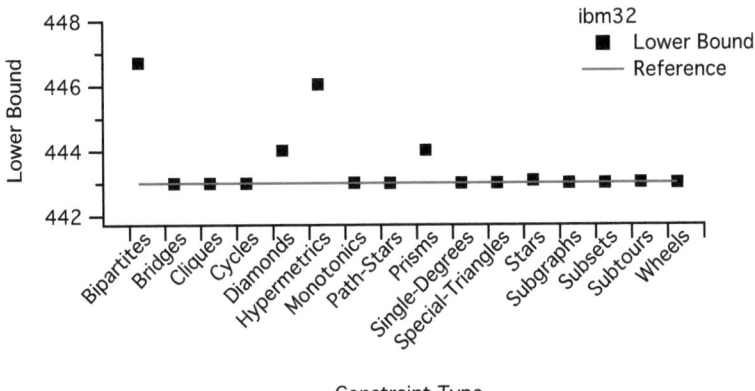

(a) Strength of the constraints for ibm32.

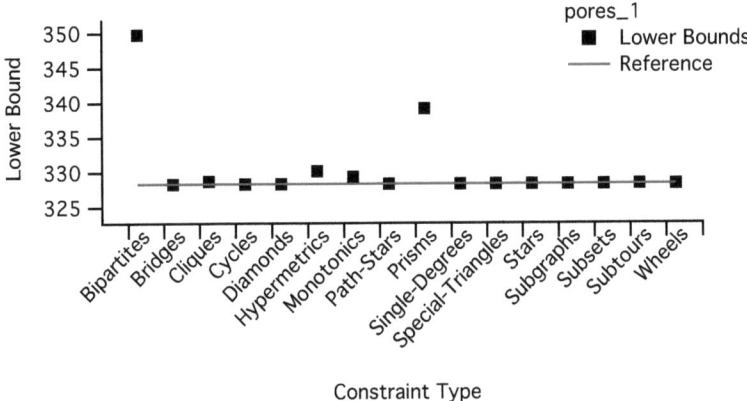

(b) Strength of the constraints for pores_1.

Figure 5.3: Strength of all types of constraints for the test instances ibm32 and pores_1.

Table 5.14: Number of LB improvements and largest difference between LB and UB.

Constraint Name	#LB improv. Total	> 5%	Largest diff. C betw. LB and UB A	B	C	Instance
Bipartites	11	3	83.61%	93.27%	9.66%	bcsstk01
Bridges	7	0	83.61%	87.64%	4.03%	bcsstk01
Cliques	8	1	83.61%	88.12%	4.51%	bcsstk01
Cycles	8	1	83.61%	95.71%	12.1%	bcsstk01
Diamonds	9	0	83.61%	86.29%	2.68%	bcsstk01
Hypermetrics	9	2	83.61%	94.01%	10.4%	bcsstk01
Monotonics	9	2	83.61%	98.24%	14.63%	bcsstk01
Path-Stars	7	2	81.42%	98.82%	17.4%	bcspwr02
Prisms	8	1	85.51%	88.32%	2.8%	pores_1
Single-Degrees	6	0	09.79%	99.35%	0.44%	can_62
Sparser-Stars	7	2	83.61%	89.8	6.19%	bcsstk01
Special-Triangles	7	3	83.61%	89.81%	6.19%	bcsstk01
Stars	7	0	83.61%	86.61%	3%	bcsstk01
Subgraphs	8	2	81.42%	89.51%	8.09%	bcspwr02
Subsets	7	0	97.04%	99.45%	2.41%	curtis54
Subtours	7	1	83.61%	89.8	6.2%	bcsstk01
Wheels	9	0	97.04%	99.59%	2.55%	curtis54

EVALUATION II

For this evaluation, we consider *all* instances. Here, we distinguish between the total LB improvements and improvements above 5%. The second and third column of Table 5.14 show that in most cases the LB improvement is below 5%. As we want our system to contain only strong constraints, we use the "> 5%" column for further conclusions.

To measure the impact of the improvement, we consider the largest gap between the LB with and without the constraint, compared to the UB. That means we define

$$A := \frac{LB_{constr.}}{UB} \times 100\% \quad \text{and} \quad B := \frac{LB_{default}}{UB} \times 100\%$$

and set $C := A - B$. Now we choose the largest difference C for all test instances and present the results in column four, five, and six of Table 5.14. To make the results more comprehensible, we give the name of the test instance for which the largest difference was obtained in column seven. The differences C lie between 2% and 17%. We rank the constraints with respect to the difference C. In Table 5.15 we present this ranking (top to bottom). Additionally, we have sorted the constraints from left to right depending on their number of LB improvements above 5%. The first thing to note is that there is indeed a correlation between both criteria. Those cuts which rarely or never cause a LB improvement have a very low impact on the LB. On the other hand, we see that constraint types with a strong effect on the LB have at least two LB improvements above 5%. We therefore choose all constraints with a high C value (above 6%) and with two or more LB improvements. They are shown in Table 5.16. Again, note that these results are based on the interpretation of *all* our Harwell-Boeing instances and are not as reliable as the results presented in Table 5.13. This is because of the possible LB decrease during pricing. Comparing Table 5.13 and 5.16, we see several similarities displayed in Table 5.17. We will separate all cuts of column three. As reliable tests can only be made for

5.6. CONSTRAINTS

Table 5.15: Ranking of the constraints obtained by evaluation II.

	LB improvement > 5%			
3×	2×	1×	0×	Difference C
	Path-Stars			17.4
	Monotonics			14.63
		Cycles		12.1
	Hypermetrics			10.4
Bipartites				9.66
	Subgraphs			8.09
		Subtours		6.2
Special-Triangles				6.19
	Sparser-Stars			6.19
		Cliques		4.51
			Bridges	4.03
			Stars	3
		Prisms		2.8
			Diamonds	2.68
			Wheels	2.55
			Subsets	2.41
			Single-Degrees	0.44

Table 5.16: Constraints chosen by evaluation II.

Path-Stars
Monotonics
Cycles
Hypermetrics
Bipartites
Subgraphs
Subtours
Special-Triangles
Sparser-Stars

Table 5.17: Comparison of evaluation I and evaluation II.

Only in Table 5.13 (I)	Only in Table 5.16 (II)	In Tables 5.13 and 5.16 (I & II)
Cliques	Path-Stars	Bipartites
Diamonds	Sparser-Stars	Hypermetrics
Prisms	Special-Triangles	Monotonics
Single-Degrees		Subgraphs
Stars		Subtours
Wheels		Cycles

Table 5.18: Exact versus heuristic separation of the triangle inequalities.

Instance			Exact		Heuristic	
Name	n	m	LB	Time	LB	Time
`bcspwr01`	39	46	88.9	4537	88.9	5543
`can_24`	24	68	200.63	55	200.63	64
`ibm32`	32	90	443.01	1268	443.01	1211
`pores_1`	30	103	328.37	901	328.37	604

the smallest four instances, we do not make any further tests about other additional constraint types. Therefore, the chosen types of constraints are the ones shown in column three of Table 5.17. The ordering in which they are mentioned corresponds to the number of total improvements for the four smallest instances, compare list Table 5.13.

Before we turn our attention to different separation possibilities, we consider two questions concerning single constraint types.

5.6.2 Exact versus Heuristic Separation of the Triangle Inequalities

For the triangle inequalities, an exact and a heuristical separation algorithm is realized. In column four and five of Table 5.18, the results obtained by the exact separation routine are presented. In column six and seven, the results of the heuristical separation are displayed. Table 5.18 shows that both separation variants are the fastest for two instances. We therefore consider the median of the relative improvement, which is 15% for the case that the heuristic separation is faster and 18% in the other case. Hence, the triangle inequalities are separated heuristically.

5.6.3 Different Subgraph Sizes

For the subgraph constraint, we have to determine a suitable size of the subgraph. We therefore test different sizes of G' and present the results of the sizes 6, 7, and 8 in Table 5.19. Table 5.19 shows that, with the only exception of instance `ibm32`, the subgraphs of size 8 lead to the highest running times. For `bcspwr01` and `can_24`, the fastest setting is the subgraph of size 6, whereas it is the subgraph of size 7 for `pores_1`. We therefore choose the setting in which the subgraphs have size 6.

5.7. Separation

Table 5.19: Comparison of different subgraph sizes.

Instance			$|G'| = 6$		$|G'| = 7$		$|G'| = 8$	
Name	n	m	LB	Time	LB	Time	LB	Time
bcspwr01	39	46	88.9	9120	88.9	9684	88.9	12585
can_24	24	68	201.52	65	201.5	70	201.5	91
ibm32	32	90	443.01	2379	443.01	2127	443.01	2061
pores_1	30	103	328.37	818	328.37	691	328.37	842

Table 5.20: Cumulative strength of chosen constraint types.

Instance			Reference		Best constraints	
Name	n	m	LB	Time	LB	Time
bcspwr01	39	46	88.9	3837	106.8	limit
can_24	24	68	200.63	55	203.9	2337
ibm32	32	90	443.01	774	464	limit
pores_1	30	103	328.37	409	351.03	35434

5.6.4 Combination of Best Constraints

This subsection is closed with the results obtained by the separation of all constraints presented in column three of Table 5.17. The fourth and fifth column of Table 5.20 show the LB and time to compute this LB obtained with the best setting we could find until now, compare Table 5.9. It can be seen that the chosen constraints have a large cumulative impact on the LB. Unfortunately, this is paid with a high price with respect to the running time of the algorithm.

5.7 Separation

We will now test all parameters concerning the separation of the chosen constraints.

5.7.1 Random versus Smart Separation

In the following we test two general methods to speed-up the separation. In column two and three of Table 5.21, we give the results of the current reference LB and time obtained by the separation of all chosen constraints. We compare these results with the random separation in column four and five, see Section 4.3.3 for explanations. In columns six and seven, this is followed by the results of the smart separation, and in columns eight and nine by the combination of both methods. Table 5.21 shows that we can fasten the running times by applying the general separation procedures. We therefore choose both methods, as only for this setting three of the RBs were computed within the time limit.

5.7.2 Different Maximal Numbers of Separated Constraints per Iteration

We now consider different maximal numbers of separated constraints per iteration. Up to now, we separated 250 cuts per iteration of each constraint type. In columns four and five of Table 5.22, we test a small number, whereas columns eight and nine show the results for 400 cuts per iteration. Again, we can easily identify the fastest setting of Table 5.22, which is to separate 100 per iteration: This is because it it the fastest setting for can_24 and pores_1.

Table 5.21: Random versus smart separation.

	Reference		Random		Smart		Both	
Instance	LB	Time	LB	Time	LB	Time	LB	Time
bcspwr01	106.8	limit	106.72	limit	105.68	limit	106.85	limit
can_24	203.9	2337	203.85	1748	203.44	2433	203.71	3088
ibm32	464	limit	463	limit	458.94	limit	463.26	84696
pores_1	351.03	35434	350.7	44373	343.15	49685	351.02	44284

Table 5.22: Comparison of different maximal numbers of separated constraints per iteration.

Instance			100 p. iter.		250 p. iter.		400 p. iter.	
Name	n	m	LB	Time	LB	Time	LB	Time
bcspwr01	39	46	105.56	limit	106.85	limit	105.87	limit
can_24	24	68	203.78	1534	203.71	3088	203.86	2595
ibm32	32	90	463.27	43702	463.26	84696	203.86	2595
pores_1	30	103	350.79	21829	343.15	49685	350.46	33295

For `ibm32` it is faster to separate 400 per iteration and for `bcspwr01` the RB is not reached within the time limit. We therefore choose to separate 100 cuts of each constraint type per iteration.

5.7.3 Cut Selection Strategies

We now turn our attention to the cut selection strategies.

Rankings

We start with the comparison of the four different rankings explained in Section 4.3.4. In column four and five of Table 5.23, the reference results of Table 5.22, columns four and five, are presented.

5.7. SEPARATION

Table 5.23: Comparison of different rankings.

Instance			No Ranking		Angle		Distance		Random		Violation	
Name	n	m	LB	Time	LB	Time	LB	Time	LB	Time	LB	Time
bcspwr01	39	46	105.56	limit	104.6	limit	105.62	limit	104.78	limit	105.75	limit
can_24	24	68	203.78	1534	203.83	1392	203.74	1950	203.91	1571	203.49	1822
ibm32	32	90	463.27	43702	463.29	79340	463.73	73635	463.27	65364	463.04	62818
pores_1	30	103	350.79	21829	350.33	36462	350.62	40874	350.56	33984	350.36	33637

Table 5.24: Comparison of different orderings of constraint separation combined with variable disjoint cut selection.

Instance			No VDCS		VDCS, forward		VDCS, reverse	
Name	n	m	LB	Time	LB	Time	LB	Time
`bcspwr01`	39	46	105.56	limit	107.59	limit	107.49	limit
`can_24`	24	68	203.78	1534	203.88	3158	203.87	2977
`ibm32`	32	90	463.27	43702	486.87	limit	488.02	limit
`pores_1`	30	103	350.79	21829	372.49	limit	372.77	limit

Table 5.23 shows the following situation: No ranking setting computes the RB of all test instances within the time limit. A closer look at the times shows that (with the only exception of `can_24` and the angle setting) the time increases for every ranking. Hence, the advantage of obtaining stronger cuts seems to be overtaken by the time consuming handling of the ranked constraints. We therefore do not apply any of the rankings but instead add all separated constraints. As a consequence, we do not test different numbers *maxConAdd* of maximal added constraints per iteration, as all separated constraints are chosen.

VARIABLE DISJOINT CUTS AND DIFFERENT ORDERINGS OF THE SEPARATION ALGORITHMS

We will now test the impact of the second cut selection strategy, the variable disjoint cut selection (VDCS) which is described in Section 4.3.4. As the order in which the constraints are separated is essential for this method, we test two different separation orders. In both cases, we start with the separation of all constraints of the IP formulation. In contrast to the separation ordering displayed in column three of Table 5.17 which we call "forward", we consider the "reverse" ordering as well.

In Table 5.24 we compare the VDCS for both separation orderings. The result show that using the variable disjoint cut selection strategy is very time consuming, independent from the separation order of the constraints. We therefore decide to run the branch-and-cut-and-price algorithm without this cut selection strategy.

5.7.4 Deep Separation Variants

Modifications of the current LP solution are explained in Section 4.3.3. In Table 5.25 we compare the deep feasible method with the deep center modification. Different values of λ have been pre-tested for both methods. Here, we present the best results for each modification, where $\lambda = 0.5$ for the deep feasible separation and $\lambda = 0.1$ in the other case.

The results of Table 5.25 show the following: With the only exception of `bcspwr01`, the fastest running times are obtained with the deep feasible separation. We therefore choose to use it in the best setting.

BEST SETTING

At the end of all tests, we summarize the best parameter setting of our algorithm.

- The IP constraints (2.2) to (2.7) are separated.
- Triangle inequalities are separated heuristically.
- Implicit use of y-variables.
- Primal simplex is used for solving the LP relaxation.
- A feasible start solution π is computed with the multi-start local search routine.

Table 5.25: Comparison of deep separation strategies.

Instance			Reference		Deep Feasible		Deep Center	
Name	n	m	LB	Time	LB	Time	LB	Time
bcspwr01	39	46	105.56	limit	103.91	limit	107.87	limit
can_24	24	68	203.78	1534	203.86	1080	210	limit
ibm32	32	90	463.27	43702	462.36	30677	492.93	limit
pores_1	30	103	350.79	21829	351.02	15340	382.95	limit

- Start variables are all variables $d_{ijk'}$ with $|\pi(i) - \pi(j)| = k'$ for $|k' - k| \leq 1$.
- At most 200 variables were priced-in per call.
- We use a combined smart and random pricing strategy.
- Additional pricing steps are performed after every fifth solved LP.
- Separation is done in a smart and random way.
- At most 100 cuts per iteration are generated of each constraint type.
- All generated cuts are added to the LP.
- No ranking is used.
- We do not apply the variable disjoint cut selection strategy.
- The feasible deep separation strategy is applied.
- Limit for the running time is 24 hours (86 400 seconds).

Provided with the best choice of all parameters of our algorithm, we tested larger instances from the Petit test set as well. But since it takes about 12 hours to solve one single LP, we did not pursue using our algorithm for the optimal solution of larger benchmark instances of the MinLA.

5.8 Improvement Heuristics

We continue with the comparison of the improvement heuristics described in Section 4.4 on page 81. Table 5.26 presents the name of the instances and their UB obtained by the multi-start local search routine described in Section 4.2.2 on page 75. Caprara et al. [7] showed that for most benchmark instances, the best-known solutions are very close to the optimum. Hence, the UBs of the MinLA are very good and in several cases even optimal. Applying the improvement heuristics, we can therefore not expect a lot of UB improvements. Therefore, we additionally show how many times the improvement heuristic reached the UB. Both numbers are given for all improvement heuristics. We present results only for the weighted edge lengths heuristic. This is because both variants of this heuristics achieve the same number of UB improvements and numbers of reached UBs. In Table 5.26 we present only those instances that have a non-zero entry for one of the heuristics. The results show that the idea of the longest distance heuristic reflects the structure of an optimal solution.

Table 5.26: Number of UB improvements and number of reached UBs shown for the different improvement heuristics.

Instance		Dist. to Border		Longest Dist.		Edge Lengths	
Name	UB	Impr.	Reached	Impr.	Reached	Impr.	Reached
can_62	222	0	0	0	3	0	0
curtis54	512	0	0	0	1	0	0
ibm32	493	0	0	0	8	0	0
gd95c	506	0	0	0	28	0	0

5.9 Branching

We now turn our attention to the test of all branch rules described in Section 4.5 on page 83 using the best setting of our algorithm. As test instances we select the instances can_24, ibm32 and pores_1, as their RB can be computed within the time limit of 24 hours. In addition, we generate some smaller test instances to have a greater variety for the tests of our branch rules.

5.9.1 Small Test Instances

We construct a set of small instances that have a challenging structure and can be solved with our algorithm within the time limit. The first type of instances is inspired by the graph drawing instances: Parts of these graphs result in the small test instances centerOf5c, mesh3x3, mesh4x3, and partOf6c_1. The second type of test instances n10_p0.20, n10_p0.30, and n10_p0.40 is generated randomly in the following way. The number of nodes n and the density p is chosen. Then we randomly generate edges between the nodes, where each edge exists with the probability p. If the resulting graph is connected, we choose it.

5.9.2 Branch on Deg-Big Constraints

Running the algorithm for all branch rules (including the different possibilities to choose the branch variables/structure), we face two problems. Both have their origin in the interface between the LP solver CPLEX and the branch-and-cut framework ABACUS. Since the latest software version ABACUS 3.0 cannot handle pricing yet, we use ABACUS 2.4. As a drawback, we have two types of mistakes: The first type is a CPLEX return value unknown to ABACUS. The second problem is the same mistake in the context of pricing. For nearly all instances and branch rules, either one of these mistakes occurs, or the time limit is reached before the optimum is computed. If this is the case, more than 10 000 sub problems are generated. For some combinations of instances and branch rules, we obtain the optimal MinLA value within the time limit. This occurs most often if we branch on deg-big constraints. In Table 5.27 we present results obtained by this branch rule. The first three columns show the instances, their number n of nodes and number m of edges. We then present the RB and the time in seconds needed to compute it. In column five we show the optimal MinLA value and continue with the time, the number of LPs and the number of sub problems necessary to compute the optimum. The results in Table 5.27 show that the RB can be computed very fast for all new instances. Furthermore, the RB is very close to the optimal MinLA value. Nevertheless, the results show that branching with this modeling approach is not as useful as expected. The theoretical advantages did not show a satisfying impact during practical use within our branch-and-cut-and-price algorithm.

Table 5.27: Branch on deg-big constraints.

Name	n	m	RB	Time$_{RB}$	Opt	Time$_{opt}$	#LPs$_{opt}$	#Subs$_{opt}$
centerOf5c	9	17	33.71	0.41	34	111	2583	793
mesh3x3	9	12	22.83	1.22	24	301	5955	2273
mesh4x3	14	21	43.16	13.84	46	limit	8850	5839
partOf6c_1	12	16	26.27	3.31	27	7370	22989	6362
n10_p0.20	10	12	18	0.42	18	497	6611	1837
n10_p0.30	10	14	25.48	0.88	27	2100	20872	7067
n10_p0.40	10	15	26.91	0.66	29	4700	38900	13149
can_24	24	68	203.86	1080	210	limit	1463	264
ibm32	32	90	462.36	30677	485	limit	465	1
pores_1	30	103	351.02	15340	383	limit	366	9

5.10 Sparse 0/1 Model

Having sorted out the best set of setting of our algorithm for the complete model, we now investigate the algorithm for the sparse problem formulation. We start with the modified system corresponding to the best setting of the complete problem formulation, compare Section 2.2.1 on page 40. The LB and time to compute it with the sparse setting is shown in columns two and three of Table 5.28. In the next two columns, we present the results for the setting in which transitive variables are added, recall Section 2.2.2 on page 40. The LB and time to obtain it with additional variables that "rescue" some of the the-longer-the-rarer equations are shown in columns six and seven. Then both variable types are added, which leads to the results presented in columns eight and nine. The results obtained with the shortest path strengthening, explained in Section 2.2.3 on page 41, are shown in column ten and eleven, whereas all ways to strengthen the sparse problem formulation are switched on for the results in the last two columns.

Table 5.28: Comparison of different ways to strengthen the sparse problem formulation.

Instance	Sparse		Trans. Var.		Rarer Eq.		Both Vars		Short.Paths		All	
	LB	Time	LB	Time	LB	Time	LB	Time	LB	Time	LB	Time
bcspwr01	62	1	61.67	limit	62	1	61.75	2	81.9	1	61.75	3
bcspwr02	86	1	87.13	8	86	1	87.13	8	120.22	1	87.13	13
bcsstk01	437	105	517.14	limit	437	123	514.94	limit	437	90	694.61	1244
can_24	144	2	150.85	limit	144	3	151.36	limit	144	2	175.13	12
can_61	816	343	903.1	limit	816	252	889.26	limit	816	336	1032.22	3756
can_62	104	2	114.72	limit	104	4	114.72	84	155.73	5	114.72	78
curtis54	256.56	34	279.95	limit	256.56	35	275.38	limit	346.56	39	280.05	189
ibm32	192.5	8	219.52	limit	192.5	8	218.42	limit	321.67	11	329.27	70
impcol_b	979	646	1186	limit	979	556	1190.45	limit	1154.05	898	1382.19	6205
pores_1	245	5	269.02	limit	245	7	268.73	limit	260.02	7	298.61	45
will57	258.81	30	-	-	258.81	31	270.15	limit	258.81	22	270.21	117
gd95c	316.9	57	359.48	limit	316.9	67	-	-	316.9	52	390.8	698
gd96b	1199	2247	1199	limit	1199	2227	1199	2173	1224.81	5163	1224.99	3760
gd96c	194	44	208.17	limit	194	44	204.12	limit	232	58	246.61	357
gd96d	921	15	1087.67	limit	921	1908	1073.86	limit	1588.63	3239	1353.1	limit

5.10. SPARSE 0/1 MODEL

The entries in Table 5.28 marked with "-" indicate that the algorithm stopped with one of the problems concerning the interface between ABACUS and CPLEX, recall Section 5.9.2. Comparing the sparse setting to the additional transitive variables setting, we observe an increase of the LB in nearly all cases. (Note: As additional variables change the structure of the constraint system, the LB can possibly be worse than in the default setting. An example is the instance `bcspwr01`.) The running time increases enormously using the transitive variables. The same situation holds if we compare the sparse setting with the additional variables of both types. In contrast to these settings, the additional "the-longer-the-rarer" variables have no impact on the LB. The running time decreases in most cases, whereas it increases for the test instances `can_61`, and `impcol_b`.

The shortest path strengthening increases the LB significantly except for the instances `bcsstk01`, `can_24`, `can_61`, `will57` and `gd95c` for which the LB remains unchanged. In case of `bcsstk01`, `will57` and `gd95c` the running time can be decreased, whereas for all other instances the running time of the shortest path strengthening is slower than in the default sparse setting. This is because the constraints of the shortest path strengthening have much more non-zero coefficients than the original ones. The setting presented in the last two columns is by far the best: With the only exception of `bcspwr01`, the LB is increased and in nearly all cases the improvement is significant. Although the running time is increased, the RB can be computed within the time limit for all instances but `gd96d`. We will therefore call this last setting the **enriched sparse** problem formulation.

5.10.1 COMPLETE VERSUS SPARSE PROBLEM FORMULATION

We will now compare the complete problem formulation to the sparse and the enriched sparse problem formulation. Table 5.29 shows the results obtained by the complete and the default sparse setting. Furthermore, the last two columns present the results for the enriched sparse setting. Due to pricing, we have to be careful interpreting the LB of those instances, for wich the time limit is reached before the RB is computed: Every pricing-in of variables reduces the LB.

The first thing to be noticed is the huge difference of the running times. If we consider the LB, the situation is very different. In several cases the LB of the enriched sparse version is half the LB of the complete version. Hence, enriching the sparse problem formulation, the great difference in strength to the complete problem formulation cannot be decreased significantly. We therefore accept the higher running time in order to compute better LBs.

Table 5.29: Complete versus sparse problem formulation. (LBs marked with "*" are above the UB because of pricing, see Section 5.5.1 on page 95 for details.)

Instance			Complete		Sparse		Enriched Sparse	
Name	n	m	LB	Time	LB	Time	LB	Time
bcspwr01	39	46	103.91	limit	62	1	61.75	3
bcspwr02	49	59	172.42	limit	86	1	87.13	13
bcsstk01	48	176	1151.62	limit	437	105	694.61	1244
can_24	24	68	203.86	1080	144	2	175.13	12
can_61	61	248	1137	limit	816	343	1032.22	3756
can_62	62	78	221.99	limit	104	2	114.72	78
curtis54	54	124	511.91	limit	256.56	34	280.05	189
ibm32	32	90	462.36	30677	192.5	8	329.27	70
impcol_b	59	281	2357.81	limit	979	646	1382.19	6205
pores_1	30	103	351.02	15340	245	5	298.61	45
will57	57	127	352	limit	258.81	30	270.21	117
gd95c	62	144	559*	limit	316.9	57	390.8	698
gd96b	111	193	1619.73*	limit	1199	2247	1224.99	3760
gd96c	65	125	525.31*	limit	194	44	246.61	357
gd96d	180	228	2494.67*	limit	921	15	1351.1	limit

5.11 Final Comparisons

We conclude this thesis with the comparison of our best results with those obtained from using other models of the MinLA.

Table 5.30 starts with the instance name and the UB obtained by the multi-start local search routine described in Section 4.2.2. All times in the following columns are given in seconds. Columns three and four show the LB and time for the betweenness approach of Caprara *et al.* [8], realized by Schwarz [113]. Recall Section 1.5.2 on page 28 for a short description of this modeling. The next two columns present the results obtained by the combined approach of Caprara *et al.* [7], see Section 1.5.2 on page 28. The instances marked with "-" have not been considered in [7]. In columns seven and eight, we display our best result, whereas in the last column a decreased LB is shown. Such a decrease of the LB is possible whenever the RB is not yet reached: With every pricing-in of variables, the LB decreases, recall Section 5.5.1 on page 95 for a detailed explanation. Only for the instances can_24, ibm32 and pores_1 the RB is reached within the time limit and does not decrease anymore. The corresponding "LB−5%" enries are therefore marked with "×". As the LB decrease of these smallest instances lies between 5% and 16% (see Section 5.5.1), we expect the LB of all other instances to decrease by the same amount.

With the only exception of gd96d, the best LBs are obtained by [113]. For the test instances can_24, ibm32 and pores_1, our LBs are below those of [113]. Nevertheless, it is higher than the LB of [7] for the test instance can_24. For all instances, the running times of [113] and [7] are much faster than ours. Considering the instances for which the LBs are not reliable, we suggest to compare our LB decreased by 5% with the LBs of [113] and [7]. With the only exception of can_61, our decreased LBs are above the ones computed by [7]. For the instances can_62, curtis54, impcol_b and will57 the decreased LB is even higher than the LB of [113].

We conclude that our modeling approach does not match the high expectations that were motivated by several theoretical advantages of the binary distance modeling. For most of the MinLA benchmark instances, the computational strength of our modeling compared to the integral distance model, recall Section 5.3 on page 146, seems to be undone by the large number (n^3) of variables. We therefore suggest pursuing other modeling approaches for solving MinLA problems to optimality.

Table 5.30: Comparison of lower bounds for different modeling approaches. (Time limit is 86 400 seconds. LBs marked with "*" are above the UB because of pricing, see Section 5.5.1 on page 95 for details.)

Instance		Reference [113]		Reference [7]		Binary distance model		
Name	UB	LB	Time	LB	Time	LB	Time	LB–5%
bcspwr01	109	106	5	91	0.7	103.91	limit	98.71
bcspwr02	173	166	11	144	1.8	172.93	limit	164.28
bcsstk01	1152	1132	14288	972	3848.1	1151.62	limit	1094.04
can_24	210	210	4	203	2.8	203.86	1080	×
can_61	1137	1137	561	1119	538	1137	limit	1080.15
can_62	222	210	59	187	4.2	221.99	limit	210.88
curtis54	512	454	88	-	-	511.91	limit	486.31
ibm32	493	485	606	-	-	462.36	30677	×
impcol_b	2358	2060.5	limit	-	-	2357.81	limit	2239.92
pores_1	383	383	16	-	-	351.02	15340	×
will57	352	334	92	-	-	352	limit	334.3
gd95c	506	506	101	443	68.3	559*	limit	531.05
gd96b	1416	1403.7	limit	1281	9.5	1619.73*	limit	1538.74
gd96c	519	519	2178	402	218.1	525.31*	limit	499.04
gd96d	2289	1612.2	limit	2021	1642.2	2494.67*	limit	2369.94

Appendix A

Unbounded Edges of Q_n

Table A.1: Unbounded edges of Q_n.

n	Slopes	Matrix	
4	2	$C^{1001} :=$	$\begin{pmatrix} 0 & 1 & -2 & 1 \\ 1 & 0 & 3 & -2 \\ -2 & 3 & 0 & 1 \\ 1 & -2 & 1 & 0 \end{pmatrix}$
5	2	$C^{11011} :=$	$\begin{pmatrix} 0 & 8 & -6 & -1 & -1 \\ 8 & 0 & 2 & 9 & -3 \\ -6 & 2 & 0 & 5 & -7 \\ -1 & 9 & 5 & 0 & 11 \\ -1 & -3 & -7 & 11 & 0 \end{pmatrix}$
5	3	$C^{10110} :=$	$\begin{pmatrix} 0 & 2 & 2 & 1 & -3 \\ 2 & 0 & 0 & -2 & 2 \\ -2 & 0 & 0 & 2 & 0 \\ 1 & -2 & 2 & 0 & 1 \\ -3 & 2 & 0 & 1 & 0 \end{pmatrix}$
5	3	$C^{10010} :=$	$\begin{pmatrix} 0 & 2 & -2 & 2 & -2 \\ 2 & 0 & 4 & -3 & 1 \\ -2 & 4 & 0 & 1 & 1 \\ 2 & -3 & 1 & 0 & 1 \\ -2 & 1 & 1 & 1 & 0 \end{pmatrix}$
5	4	$C^{10101} :=$	$\begin{pmatrix} 0 & 0 & 3 & -2 & -1 \\ 0 & 0 & 1 & 1 & -2 \\ 3 & 1 & 0 & 1 & 3 \\ -2 & 1 & 1 & 0 & 0 \\ -1 & -2 & 3 & 0 & 0 \end{pmatrix}$
6	5	$C^{101010} :=$	$\begin{pmatrix} 0 & 0 & 1 & -1 & 0 & 0 \\ 0 & 0 & 1 & 1 & -2 & 0 \\ 1 & 1 & 0 & 1 & 3 & -2 \\ -1 & 1 & 1 & 0 & 0 & 1 \\ 0 & -2 & 3 & 0 & 0 & 1 \\ 0 & 0 & -2 & 1 & 1 & 0 \end{pmatrix}$

List of Algorithms

1	Simplex Algorithm with an Initial Basis	10
2	InfeasibleLP()	15
3	InitMakeFeas(A_{ℓ}, b_{ℓ}, c)	16
4	MakeFeasible(A, b, c, x^*, y^*, B)	17
5	Embedability Test	49
6	Determine if G is k-spanning	50
7	Determine all possible embedings of G	50
8	Determine all possible embedings of $u_1, \ldots, u_k$	51
9	Check existence of continuation	51
10	VariableDisjointCutSelection()	80
11	DistanceToTheBorderHeuristic()	81
12	LongestDistancesHeuristic()	82
13	EdgeLengthHeuristic()	83
14	Set Relative Distances of Two Nodes	84
15	Choice of Suitable Branching Structure G'	87

List of Figures

1	Flowchart of a branch-and-bound algorithm.	12
2	Flowchart of a branch-and-cut-and-price algorithm	14
1.1	An example graph and its MinLA presentation.	19
1.2	Several graphs and their MinLA presentation.	21
1.3	Example for the bipartite inequalities of the MinLA considered in this thesis.	26
2.1	Several forbidden subgraphs of bridge type.	35
2.2	Several forbidden subgraphs of path-star type.	35
2.3	Example of a 4-bridge within a current LP solution.	36
2.4	Diamond and its MinLA representation.	38
2.5	Cycle-stars and their MinLA representation.	38
2.6	3-cycle-with-legs and their MinLA representation.	39
2.7	4-cycle-with-legs and their MinLA representation.	39
2.8	Additional variables of transitive type.	41
3.1	Convex set Q_3 in $M(L^3)$ dispayed from three perspectives.	54
4.1	Example for 2-exchange.	76
4.2	Intervals for branching on the deg-big constraint.	84
5.1	Behavior of the LB in a branch-and-cut-and-price algorithm.	96
5.2	Strength of all types of constraints for the test instances `bcspwr01` and `can_24`.	104
5.3	Strength of all types of constraints for the test instances `ibm32` and `pores_1`.	105

LIST OF TABLES

0.1	Correspondence of the primal and dual linear program.	8
1.1	Time complexity of algorithms that solve the MinLA of certain graphs to optimality.	22
1.2	Overview of different meta heuristics for the MinLA.	23
1.3	Overview of different MinLA specific heuristics.	23
3.1	Are there cut cone facets that are facets of P_n?	58
3.2	List of all sets with two slopes (up to complement).	64
3.3	List of all sets with three slopes (up to complement).	65
5.1	Properties of the test instances.	90
5.2	Improvement of binary distance model compared to the integral distance modeling approach	92
5.3	Explicit versus implicit use of y-variables.	94
5.4	Comparison of different LP solver settings.	94
5.5	Comparison of different start sets of variables	96
5.6	Comparison of different sizes of the start set of variables.	97
5.7	Comparison of different maximal numbers of priced-in variables per call.	98
5.8	Comparison of different pricing strategies.	98
5.9	Comparison of different frequencies of additional pricing steps.	98
5.10	Strength of constraint types (A–H).	100
5.11	Strength of constraint types (I–Sp).	101
5.12	Strength of constraint types (St–Z).	102
5.13	Constraints chosen by evaluation I.	103
5.14	Number of LB improvements and largest difference between LB and UB.	106
5.15	Ranking of the constraints obtained by evaluation II.	107
5.16	Constraints chosen by evaluation II.	107
5.17	Comparison of evaluation I and evaluation II.	108
5.18	Exact versus heuristic separation of the triangle inequalities.	108
5.19	Comparison of different subgraph sizes.	109
5.20	Cumulative strength of chosen constraint types.	109
5.21	Random versus smart separation.	110
5.22	Comparison of different maximal numbers of separated constraints per iteration.	110
5.23	Comparison of different rankings.	111
5.24	Comparison of different orderings of constraint separation combined with variable disjoint cut selection.	112
5.25	Comparison of deep separation strategies.	113
5.26	Number of UB improvements and number of reached UBs shown for the different improvement heuristics.	114

5.27	Branch on deg-big constraints.	115
5.28	Comparison of different ways to strengthen the sparse problem formulation.	116
5.29	Complete versus sparse problem formulation.	118
5.30	Comparison of lower bounds for different modeling approaches.	120
A.1	Unbounded edges of Q_n.	124

REFERENCES

[1] J. Díaz, J. Petit i Silvestre & P. Spirakis. *Heuristics for the MinLA problem: Some theoretical and empirical considerations.* 1998. pages 1 and 23.

[2] J. Petit i Silvestre. *Experiments on the minimum linear arrangement problem.* Technical report LSI-01-7-R, Universitat Politècnica de Catalunya, Departament de Llenguatges i Sistemes Informatics, Spain, 2001. pages 1, 23, and 89.

[3] Y. Koren & D. Harel. *A multi-scale algorithm for the linear arrangement problem.* Lecture Notes In Computer Science; Revised Papers from the 28th International Workshop on Graph-Theoretic Concepts in Computer Science, 2573, 296–309, 2002. pages 1, 20, and 23.

[4] I. Safro, D. Ron & A. Brandt. *Graph minimum linear arrangement by multilevel weighted edge contractions.* Journal of Algorithms, 60(1), 24–41, 2006. doi:10.1016/j.jalgor.2004.10.004. pages 1 and 23.

[5] G. Even, J. S. Naor, S. Rao & B. Schieber. *Divide-and-conquer approximation algorithms via spreading metrics.* Journal of the ACM, 47(4), 585–616, 2000. doi:10.1145/347476.347478. pages 1 and 27.

[6] W. Liu & A. Vannelli. *Generating lower bounds for the linear arrangement problem.* Discrete Applied Mathematics, 59(2), 137–151, 1995. doi:10.1016/0166-218X(93)E0168-X. pages 1, 21, 24, and 28.

[7] A. Caprara, A. N. Letchford & J.-J. Salazar-Gonzales. *Decorous lower bounds for minimum linear arrangement.* INFORMS Journal on Computing, 2010. Accepted (February 2010). pages 1, 26, 28, 29, 41, 49, 77, 97, 113, 119, and 120.

[8] A. Caprara, M. Jung, M. Oswald, G. Reinelt & E. Traversi. *A betweenness approach for solving the linear arrangement problem,* 2010. In preparation. pages 1, 29, and 119.

[9] G. Reinelt. *Effiziente Algorithmen I,* 2001. Lecture notes. Institut für Informatik, Ruprecht-Karls-Universität, Heidelberg. page 3.

[10] R. Diestel. *Graph theory,* volume 173 of *Graduate Texts in Mathematics.* Springer-Verlag, 1997. ISBN 978-0-387-98211-3. page 3.

[11] G. Galambos. *Graph theory,* 2005. Lecture notes. Institut für Informatik, Ruprecht-Karls-Universität, Heidelberg. page 3.

[12] A. Schrijver. *Combinatorial optimization: Polyhedra and efficiency,* volume 24 of *Algorithms and Combinatorics.* Springer, 2003. pages 3 and 7.

[13] P. Assouad. *Plongements isométriques dans L^1: Aspect analytique.* In *Initiation Seminar on Analysis: G. Choquet, M. Rogalski, J. Saint-Raymond, 19th Year: 1979/1980,* volume 14 of *Mathematical Publications of the Université Pierre et Marie Curie,* pp. 1–23. Univ. Paris VI, Paris, 1980. page 5.

[14] F. Barahona & A. R. Mahjoub. *On the cut polytope.* Mathematical Programming, 36(2), 157–173, 1986. doi:10.1007/BF02592023. pages 5 and 67.

[15] G. Fischer. *Lineare Algebra.* Vieweg, 13th edition, 2005. page 5.

[16] F. Lorenz. *Lineare Algebra I+II.* Wissenschaftsverlag, 1992. page 5.

[17] G. M. Ziegler. *Lecture on polytopes.* Graduate Texts in Mathematics. Springer, 1995. pages 6, 7, 8, 43, 44, and 45.

[18] W. J. Cook, W. H. Cunningham, W. R. Pulleyblank & A. Schrijver. *Combinatorial Optimization.* Wiley-Interscience Series in Discrete Mathematics and Optimization. Johnson Wiley and Sons, Inc., 1997. ISBN 978-0-471-55894-1. pages 6, 7, and 9.

[19] Y. Pochet & L. A. Wolsey. *Production planning by mixed integer programming.* Springer Series in Operations Research

and Financial Engeneering. Springer, 2006. ISBN 978-0387299594. pages 6 and 7.

[20] M. GRÖTSCHEL, L. LOVÁSZ & A. SCHRIJVER. *Geometric algorithms and combinatorial optimization*. Springer. Berlin, 1988. ISBN 978-0387136240. page 7.

[21] L. G. KHACHIYAN. A polynomial algorithm in linear programming. *Proceedings of the USSR Academy of Sciences*, 224, 1093–1096, 1979. (in russian). page 8.

[22] N. KARMARKAR. A new polynomial-time algorithm for linear programming. *Combinatorica*, 4, 373–395, 1984. doi:10.1007/BF02579150. page 8.

[23] V. CHVÁTAL. *Linear programming*. Graduate Texts in Mathematics. Springer, 2000. page 8.

[24] A. SCHRIJVER. *Theory of linear programming*. Wiley. Chichester, 1986. page 8.

[25] S. MEHROTRA. On the implementation of a primal-dual interior point method. *SIAM Journal on Optimization*, 2(4), 575–601, 1992. doi:10.1137/0802028. page 8.

[26] ILOG. *Cplex 8.1*. 2002. By ILOG S.A., 9 Rue de Verdun, 94253 Gentilly Cedex, France, http://www.ilog.com/products/cplex. pages 9, 13, and 90.

[27] KONRAD-ZUSE-ZENTRUM FÜR INFORMATIONSTECHNIK BERLIN. *The Sequential Object-oriented simPLEX class library*. 1996. http://www.zib.de/Optimization/Software/Soplex. page 9.

[28] ROLAND WUNDERLING. *Paralleler und objektorientierter Simplex-Algorithmus*. Ph.D. thesis, Technische Universität Berlin, 1996. http://www.zib.de/Publications/abstracts/TR-96-09/. page 9.

[29] G. B. DANTZIG & P. WOLFE. The decomposition algorithm for linear programs. *Econometrica*, 29(4), 767–778, 1961. page 10.

[30] J. F. BENDERS. Partitioning prodedures for solving mixed-variables programming problems. *Numerische Mathematik*, 4, 238–258, 1962. doi:10.1007/s10287-004-0020-y. page 10.

[31] M. R. GAREY & D. S. JOHNSON. *Computers and intractability, a guide to the theory of NP-completeness*. A series of books in the mathematical sciences. W. H. Freeman, 1979. ISBN 978-0716710455. page 11.

[32] G. REINELT. *The traveling salesman: Computational solutions for TSP applications*, volume 840 of *Lecture Notes in Computer Science*. Springer-Verlag, 1994. ISBN 3-540-58334-3. page 11.

[33] D. L. APPLEGATE, R. E. BIXBY & W. J. COOK. *The traveling salesman problem: A computational study*. Princeton Series in Applied Mathematics. Springer-Verlag, 1994. page 11.

[34] G. REINELT. *Effiziente Algorithmen II*, 2007. Lecture notes. Institut für Informatik, Ruprecht-Karls-Universität, Heidelberg. pages 11 and 75.

[35] S. THIENEL. *ABACUS: A Branch-And-CUt System*. Ph.D. thesis, Universtät zu Köln, 1995. pages 11, 13, and 90.

[36] A. H. LAND & A. G. DOIG. An automatic method of solving discrete programming problems. *Econometrica*, 28(3), 497–520, 1960. page 11.

[37] R. J. DAKIN. A tree-search algorithm for mixed integer programming problems. *The Computer Journal*, 8(3), 250–255, 1965. doi:10.1093/comjnl/8.3.250. page 11.

[38] UNIVERSTÄT ZU KÖLN. *ABACUS - A Branch-And-CUt System 2.4*. 2003. http://www.informatik.uni-koeln.de/abacus. page 13.

[39] M. JÜNGER, G. REINELT & S. THIENEL. Provably good solutions for the traveling salesman problem. *Mathematical Methods of Operations Research*, 40(2), 183–217, 1994. doi:10.1007/BF01432809. page 13.

[40] E. L. LAWLER. Procedure for computing the k-best solutions to discrete optimization problems and its applications to the shortest path problem. *Management Science*, 18(7), 401–405, 1972. doi:10.1287/mnsc.18.7.401. page 15.

[41] C. C. RIBEIRO, M. MINOUX & M. C. PENNA. An optimal column-generation-with-ranking algorithm for very large scale set partitioning problems in traffic assignment. *European Journal of Operational Research*, 41(2), 232–239, 1989. doi:10.1016/0377-2217(89)90389-5. page 15.

[42] D. M. RYAN & B. A. FOSTER. An integer programming approach to scheduling. A. Wren (Eds.): *Computer scheduling of public transport urban passenger vehicle and crew scheduling*. North Holland, Amsterdam, 1981. page 15.

[43] L. H. HARPER. Optimal assignments of numbers to vertices. *SIAM Journal on Applied Mathematics*, 12(1), 131–135, 1964. doi:10.1137/0112012. pages 19 and 20.

[44] M. HANAN & J. M. KURTZBERG. A review of the placement and quadratic assignment problems. *SIAM Review*, 14(2), 324–342,

1972. doi:10.1137/1014035. page 19.

[45] A. AMARAL & A. LETCHFORD. *Integer polyhedra associated with single row facility layout problems*, 2008. Submitted. pages 19, 25, 28, 42, 57, 67, 68, 71, 73, and 78.

[46] R. RAVI, A. AGRAWAL & P. KLEIN. *Ordering problems approximated: Single-processor scheduling and interval graphs connection*. Lecture Notes in Computer Science, 150, 751–762, 1991. pages 19 and 20.

[47] S. WIESBERG. *The multidimensional grid arrangement problem*. Diploma thesis, Universität Heidelberg, 2010. pages 20 and 32.

[48] R. HASSIN & S. RUBINSTEIN. *Algorithm Theory - SWAT 2000*, volume 1851 of *Lecture Notes in Computer Science*, chapter Approximation Algorithms for Maximum Linear Arrangement, pp. 633–643. Springer Berlin / Heidelberg, 2000. ISBN 978-3-540-67690-4. doi:10.1007/3-540-44985-X. page 20.

[49] G. GUTIN, A. RAFIEY, S. SZEIDER & A. YEO. *The linear arrangement problem parametrized above guaranteed value*. Theory of Computing Systems, 41(3), 521–538, 2007. doi:10.1007/s00224-007-1330-6. page 20.

[50] M. SERNA & D. M. THILIKOS. *Parameterized complexity for graph layout problems*. EATCS Bulletin, 86, 41–65, 2005. page 20.

[51] H. FERNAU. *Parameterized algorithmics: A graph-theoretic approach*. 2005. Habilitation thesis. page 20.

[52] H. FERNAU. *Parameterized algorithmics for linear arrangement problems*. Discrete Applied Mathematics, 156(17), 3166–3177, 2008. doi:10.1016/j.dam.2008.05.008. page 20.

[53] A. VANNELLI & G. S. ROWAN. *An eigenvector based approach for multistack VLSI layout*. Proceedings of the Midwest Symposium on Circuits and Systems, 29, 435–439, 1986. page 20.

[54] P. P. S. CHEN. *The entity-relationship model—toward a unified view of data*. ACM Transactions on Database Systems, 1, 9–36, 1976. doi:10.1145/320434.320440. page 20.

[55] C. GANE & T. SARSON. *Structured Systems Analysis: Tools and Techniques*. McDonnell Douglas Systems Integration Company, 1st edition, 1977. ISBN 978-0930196004. page 20.

[56] D. ADOLPHSON. *Single machine job sequencing with precedence constraints*. SIAM Journal on Computing, 6(1), 40–54, 1977. doi:10.1137/0206002. page 20.

[57] R. M. KARP. *Mapping the genome: Some combinatorial problems arising in molecular biology*. Proceedings of the Twenty-Fifth Annual ACM Symposium on Theory of Computing, pp. 278–285, 1993. doi:10.1145/167088.167170. page 20.

[58] G. MITCHISON & R. DURBIN. *Optimal numberings of an $N \times N$ array*. SIAM Journal on Algebraic and Discrete Methods, 7(4), 571–582, 1986. doi:10.1137/0607063. pages 20, 22, and 24.

[59] M. R. GAREY, D. S. JOHNSON & L. STOCKMEYER. *Some simplified NP-complete graph problems*. Theoretical Computer Science, 1(3), 237–267, 1976. doi:10.1016/0304-3975(76)90059-1. page 20.

[60] S. EVEN & Y. SHILOACH. *NP-completeness of several arrangements problems*, 1978. Technical Report, TR-43 The Technicon, Haifa. page 20.

[61] M. JUVAN & B. MOHAR. *Optimal linear labelings and eigenvalues of graphs*. Discrete Applied Mathematics, 36(2), 153–168, 1992. doi:10.1016/0166-218X(92)90229-4. pages 20, 21, 23, 24, and 29.

[62] L. H. HARPER. *Chassis layout and isoperimeter problems*, 1972. Preprint of Jet Propulsion Lab., California Institute of Pasadena. page 22.

[63] D. O. MURADYAN & T. E. PILIPOSJAN. *Minimal numberings of vertices of a rectangular lattice*. Akad. Nauk. Armjan. SRR, 1(70), 21–27, 1980. (in russian). pages 22 and 24.

[64] F. R. K. CHUNG. *Selected topics in graph theory*, volume 3, chapter 7 Labelings of graphs, pp. 151–168. Academic Press, 1988. ISBN 978-0-120-86203-0. pages 22 and 24.

[65] D. ADOLPHSON & T. C. HU. *Optimal linear ordering*. SIAM Journal on Applied Mathematics, 25(3), 403–423, 1973. doi:10.1137/0125042. pages 22 and 24.

[66] F. R. K. CHUNG. *The Theory and Applications of Graphs*, chapter Some problems and results on labelings of graphs, pp. 255–264. John Wiley and Sons, 1981. page 22.

[67] K. NAKANO. *Graph-Theoretic Concepts in Computer Science*, volume 790 of *Lecture Notes in Computer Science*, chapter Linear

layouts of generalized hypercubes, pp. 364–375. Springer Berlin / Heidelberg, 1994. ISBN 978-3-540-57899-4. doi:10.1007/3-540-57899-4. page 22.

[68] B. HENDRICKSON & R. LELAND. *Multidimensional spectral load balancing*. Technical Report SAND-93-0074, Sandia National Labs., Albuquerque, NM, USA, 1993. 6th SIAM Conf. Parallel Proc. Sci. Comput. page 22.

[69] T. EASTON, S. B. HORTON & R. G. PARKER. *A solvable case of the optimal linear arrangement problem on Halin graphs*. Congressus Numerantium, 119(3), 3–18, 1996. page 22.

[70] G. N. FREDERICKSON & S. E. HAMBRUSCH. *Planar linear arrangements of outerplanar graphs*. IEEE Transactions on Circuits and Systems, 35(3), 323–332, 1988. doi:10.1109/31.1745. page 22.

[71] I. SAFRO. *The minimum linear arrangement problem on proper interval graphs*. ArXiv Computer Science e-prints, abs/cs/0608008, 2006. page 22.

[72] J. PETIT I SILVESTRE. *Approximation heuristics and benchmarkings for the MinLA problem*. Proceedings of "Algorithms and Experiments" (ALEX98), pp. 112–128, 1998. page 23.

[73] J. PETIT I SILVESTRE. *Layout problems*. Ph.D. thesis, Universitat Polit'ecnica de Catalunya, 2001. pages 22, 23, 24, and 29.

[74] T. PORANEN. *A generic hillclimbing algorithm for the optimal linear arrangement problem*. Technical report, University of Tampere, Finland, 2002. page 23.

[75] J. PETIT I SILVESTRE. *Combining spectral sequencing with simulated annealing for the MinLA problem: Sequential and parallel heuristics*. Technical Report LSI-9746-R, Universitat Politècnica de Catalunya, Departament de Llenguatges i Sistemes Informatics, Spain, 1997. page 23.

[76] M. YAGIURA & T. IBARAKI. *The use of dynamic programming in genetic algorithms for permutation problems*. European Journal of Operational Research, 92(2), 387–401, 1996. doi:10.1016/0377-2217(94)00301-7. page 23.

[77] R. BAR-YEHUDA, G. EVEN, J. FELDMAN & J. NAOR. *Computing an optimal orientation of a balanced decomposition tree for linear arrangement problems*. Journal of Graph Algorithms and Applications, 5(4), 1–27, 2001. page 23.

[78] D. R. REEVES. *Modern heuristic techniques for combinatorial problems*. John Wiley & Sons, Inc., New York, NY, USA, 1993. ISBN 978-0-632-03238-9. page 23.

[79] E. RODRIGUEZ-TELLO, J.-K. HAO & J. TORRES-JIMENEZ. *Artificial Evolution*, volume 3871 of *Lecture Notes in Computer Science*, chapter Memetic Algorithms for the MinLA Problem, pp. 73–84. Springer Berlin / Heidelberg, 2006. ISBN 978-3-540-33589-4. doi:10.1007/11740698. page 23.

[80] E. RODRIGUEZ-TELLO, J.-K. HAO & J. TORRES-JIMENEZ. *MICAI 2005: Advances in Artificial Intelligence*, volume 3789 of *Lecture Notes in Computer Science*, chapter A Comparison of memetic recombination operators for the MinLA problem, pp. 613–622. Springer Berlin / Heidelberg, 2005. ISBN 978-3-540-29896-0. doi:10.1007/11579427. page 23.

[81] A. J. MCALLISTER. *A new heuristic algorithm for the linear arrangement problem*. Technical Report TR00_126a, Faculty of Computer Science, University of New Brunswick, Canada, 1999. page 23.

[82] L. MARRO. *A linear time implementation of profile reduction algorithms for sparse matrices*. SIAM Journal on Scientific Computing, 7(4), 1212–1231, 1986. doi:10.1137/0907082. page 23.

[83] S. B. HORTON. *The optimal linear arrangement problem: Algorithms and approximation*. Ph.D. thesis, Georgia Institute of Technology, USA, 1997. page 22.

[84] J. DÍAZ, J. PETIT I SILVESTRE & M. SERNA. *A survey on graph layout problems*. ACM Computing Surveys, 34(3), 313–356, 2002. doi:10.1145/568522.568523. page 22.

[85] N. R. DEVANUR, S. A. KHOT, R. SAKET & N. K. VISHNOI. *On the hardness of minimum linear arrangement*. 2006. Working paper, College of Computing, Georgia Tech. page 22.

[86] A. FRIEZE & R. KANNAN. *The regularity lemma and approximation schemes for dense problems*. Proceedings of the 37th Annual Symposium on Foundations of Computer Science, IEEE Computer Society, Washington, DC, USA, p. 12, 1996. doi:10.1109/SFCS.1996.548459. page 22.

[87] C. AMBÜHL, M. MASTROLILLI & O. SVENSSON. *Inapproximability results for sparsest cut, optimal linear arrangement, and presedence constrainted scheduling*. Proceedings of the 48th Annual Symposium on Foundations of Computer Science, IEEE Computer Society, Providence, RI, USA, pp. 329–337, 2007. doi:10.1109/FOCS.2007.40. page 22.

[88] S. RAO & A. W. RICHA. *New approximation techniques for some linear ordering problems*. SIAM Journal on Computing, 34(2),

388–404, 2005. doi:10.1137/S0097539702413197. page 22.

[89] C. F. BORNSTEIN & S. VEMPALA. *Flow metrics. Theoretical Computer Science*, 321(1), 13–24, 2004. doi:10.1016/j.tcs.2003.05.003. pages 22 and 27.

[90] S. ARORA, S. RAO & U. VAZIRANI. *Expander flows, geometric embeddings and graph partitioning*. Proceedings of the thirty-sixth annual ACM symposium on Theory of computing, Chicago, IL, USA, pp. 222–231, 2004. doi:10.1145/1007352.1007355. page 22.

[91] U. FEIGE & J. R. LEE. *An improved approximation ratio for the minimum linear arrangement problem. Information Processing Letters*, 101(1), 26–29, 2007. doi:10.1016/j.ipl.2006.07.009. pages 22 and 29.

[92] M. CHARIKAR, M. T. HAJIAGHAYI, H. KARLOFF & S. RAO. ℓ_2^2 *spreading metrics for vertex ordering problems*. Proceedings of the seventeenth annual ACM-SIAM symposium on Discrete algorithms, Miami, FL, USA, pp. 1018–1027, 2006. doi:10.1145/1109557.1109670. pages 22 and 29.

[93] N. R. DEVANUR, S. A. KHOT, R. SAKET & N. K. VISHNOI. *Integrality gaps for sparsest cut and minimum linear arrangement problems*. Proceedings of the thirty-eighth annual ACM symposium on Theory of computing, Seattle, WA, USA, pp. 537–546, 2006. doi:10.1145/1132516.1132594. page 22.

[94] R. E. GOMORY & T. C. HU. *Multi-terminal network flows. SIAM Journal on Applied Mathematics*, 9(4), 551–570, 1961. doi:10.1137/0109047. page 24.

[95] M. DEZA & M. LAURENT. *Geometry of Cuts and Metrics*. Springer Verlag New York, 1997. ISBN 978-3-540-61611-5. pages 25 and 29.

[96] A. LETCHFORD. *Personal communication*. 2006. pages 25 and 85.

[97] C. HELMBERG & F. RENDL. *Solving quadratic (0,1)-problems by semidefinite programming and cutting planes. Mathematical Programming*, 82(3), 291–315, 1998. doi:10.1007/BF01580072. pages 25 and 77.

[98] M. OSWALD. *Weighted consecutive ones problems*. Ph.D. thesis, Ruprecht-Karls-Universität, Heidelberg, 2003. http://www.ub.uni-heidelberg.de/archiv/3588. pages 28 and 29.

[99] A. TUCKER. *A structure theorem for the consecutive 1's property. Journal of Combinatorial Theory, Series B*, 12(2), 153–162, 1972. doi:10.1016/0095-8956(72)90019-6. page 28.

[100] C. BUCHHEIM, F. LIERS & M. OSWALD. *Experimental Algorithms*, volume 5038 of *Lecture Notes in Computer Science*, chapter A Basic Toolbox for Constrained Quadratic 0/1 Optimization, pp. 249–262. Springer Berlin / Heidelberg, 2008. ISBN 978-3-540-68548-7. doi:10.1007/978-3-540-68552-4. page 29.

[101] G. REINELT. *The linear ordering problem: Algorithms and applications*. Research and Exposition in Mathematics. Heldermann, Berlin, 1985. ISBN 978-3-885-38208-9. page 29.

[102] A. LETCHFORD, G. REINELT, H. SEITZ & D. O. THEIS. *On a class of metrics related to graph layout problems*, 2009. Submitted to Linear Algebra and its Applications, August 2008. Revised. pages 43 and 66.

[103] E. BALAS. *A linear characterization of permutation vectors*, 1975. Research Report 364. Carnegie-Mellon University, Pittburg, Pennsylvania, USA. pages 43 and 44.

[104] A. R. S. AMARAL & A. N. LETCHFORD. *Linear arrangement polytopes*, 2009. Working paper, Dept. of Management Science, Lancaster University. pages 48 and 67.

[105] A. R. S. AMARAL & A. N. LETCHFORD. *Some polyhedra associated with the linear arrangements and path metrics*, 2005. Working paper, Dept. of Management Science, Lancaster University. page 56.

[106] B. GRÜNBAUM. *Convex polytopes*. Graduate Texts in Mathematics. Springer, 2nd edition, 2003. ISBN 978-0-387-00424-2. page 57.

[107] T. CHRISTOF. *SMAPO, Libary of linear descriptions of SMAll problem instances of POlytopes in combinatorial optimization*. 1997. http://www.informatik.uni-heidelberg.de/groups/comopt/software/SMAPO. page 57.

[108] T. CHRISTOF & A. LÖBEL. *PORTA - POlyhedron Representation Transformation Algorithm*. 2009. http://www.zib.de/Optimization/Software/Porta/. page 70.

[109] G. REINELT. *Personal communication*. 2007. pages 75 and 79.

[110] B. W. KERNINGHAN & S. L. LIN. *An efficient heuristic procedure for partitioning graphs. The Bell System Technical Journal*,

49(2), 291–307, 1970. page 76.

[111] T. CHRISTOF & G. REINELT. *Algorithmic aspects of using small instance relaxations in parallel branch-and-cut*. Algorithmica, 30(4), 597–629, 2001. doi:10.1007/s00453-001-0029-3. page 80.

[112] I. DUFF, R. GRIMES & J. LEWIS. *User's guide for the Harwell-Boeing sparse matrix collection*. Technical Report TR/PA/92/86, CERFACS, Toulouse, France, 1992. http://people.sc.fsu.edu/~burkardt/pdf/hbsmc.pdf. page 89.

[113] R. SCHWARZ. *A branch-and-cut algorithm with betweenness variables for the linear arrangement problems*. Diploma thesis, Universität Heidelberg, 2010. pages 119 and 120.

Nomenclature

$[n]$	set $\{1,\ldots,n\}$
$\mathbf{0}$	appropriately sized vector consisting of ones
$\mathbf{1}$	appropriately sized vector consisting of ones
aff	affine hull
χ	incidence vector
conv	convex hull
$\delta(S)$	cut of node subset S
$\dim(P)$	dimension of polyhedron P
$\mathcal{DP}$	dual program
ℓ_1	embeddable in the real line
$\mathrm{exray}(P)$	set of all extreme rays of polyhedron P
$\mathcal{F}$	fan
$\mathcal{F}(P)$	face fan of polyhedron P
ι_n	identical permutation of size n
$\mathrm{int}(P)$	set of all interior points of polyhedron P
$\mathcal{IP}$	integer program
$(L^n)^*$	dual vector space of L^n
$\mathbb{1}$	square matrix of order n whose (k,l)-entry is 1 if $k \neq l$ and 0 otherwise
$\mathbb{M}(n,\mathbb{R})$	set of all $n \times n$ matrices with entries in $\mathbb{R}$
$\mathcal{N}(P)$	normal fan of polyhedron P
$\mathcal{A}$	linear hyperplane arrangement
$\mathcal{F}_\mathcal{A}$	fan of the linear hyperplane arrangement
$\deg(i)$	degree of node i
$\deg_{\mathrm{inDeg}(i)}$	in-degree of node i
$\deg_{\mathrm{outDeg}(i)}$	out-degree of node i
$\mathbb{0}$	all-zeros matrix not necessarily square
π	permutation
π^-	antipodal permutation
Π^{n-1}	$(n-1)$-dimensional permutahedron

$\mathcal{PP}$	primal program
rank(G)	rank of a graph
rec(P)	recession cone of polyhedron P
sign	signum
$S_0\mathbb{M}(n)$	set of all symmetric $n \times n$ matrices with zero in the diagonal
tr(A,B)	trace of matrices A and B
vert(P)	set of vertices of polyhedron P
$A \bullet B$	trace of matrices A and B
c_{ij}	weight of edge ij
D_S	cut matrix
e_k	k^{th} unit vector
E_π	permutation matrix
$G = (V,E)$	graph with vertex and edge set
ij	edge from node i to vertex j
M	mapping from $\mathbb{R}^n$ to $S_0\mathbb{M}(n)$
N_F	element of the normal fan corresponding to face F
N_π	element of the normal fan corresponding to face $\{\pi\}$
$P^\triangle$	polar of polyhedron P
$P_1 + P_2$	Minkowski sum of P_1 and P_2
$S(n)$	set of all permutations of dimension n
#LPs	number of solved linear programs
CUT$_n$	cut cone
LB	lower bound
LP	linear program
MLS	multi-start local search routine
OLA	optimal linear arrangement
RB	root bound
SA	simulated annealing
UB	upper bound

Die VDM Verlagsservicegesellschaft sucht für wissenschaftliche Verlage abgeschlossene und herausragende

Dissertationen, Habilitationen, Diplomarbeiten, Master Theses, Magisterarbeiten usw.

für die kostenlose Publikation als Fachbuch.

Sie verfügen über eine Arbeit, die hohen inhaltlichen und formalen Ansprüchen genügt, und haben Interesse an einer honorarvergüteten Publikation?

Dann senden Sie bitte erste Informationen über sich und Ihre Arbeit per Email an *info@vdm-vsg.de*.

Sie erhalten kurzfristig unser Feedback!

VDM Verlagsservicegesellschaft mbH
Dudweiler Landstr. 99
D - 66123 Saarbrücken
www.vdm-vsg.de

Telefon +49 681 3720 174
Fax   +49 681 3720 1749

Die VDM Verlagsservicegesellschaft mbH vertritt

Printed by Books on Demand GmbH, Norderstedt / Germany